The Key to
Newton's Dynamics

The Key to Newton's Dynamics

The Kepler Problem and the *Principia*

Containing an English Translation of
Sections 1, 2, and 3 of Book One from
the First (1687) Edition of Newton's
Mathematical Principles of Natural Philosophy

J. Bruce Brackenridge

with English translations from the Latin by
Mary Ann Rossi

UNIVERSITY OF CALIFORNIA PRESS
Berkeley   Los Angeles   London

LIBRARY
COLBY-SAWYER COLLEGE
NEW LONDON, NH 03257

University of California Press
Berkeley and Los Angeles, California

University of California Press
London, England

Copyright © 1995 by
The Regents of the University of California

Library of Congress Cataloging-in-Publication Data
Brackenridge, J. Bruce, 1927–
   The key to Newton's dynamics: the Kepler problem and the Principia: containing an English translation of sections 1, 2, and 3 of book one from the first (1687) edition of Newton's Mathematical principles of natural philosophy / J. Bruce Brackenridge; with English translations from the Latin by Mary Ann Rossi.
     p.   cm.
   Includes bibliographical references and index.
   ISBN 0–520–20065–9 (c: alk. paper).—ISBN 0–520–20217–1 (p: alk. paper)
   1. Kepler's laws. 2. Celestial mechanics. I. Newton, Isaac, Sir, 1642–1727. Principia. Book 1, sections 1–3. English. II. Title.
QB355.B694   1996
521'.3—dc20   95-32978
                                                                                    CIP

Printed in the United States of America

1 2 3 4 5 6 7 8 9

The paper used in this publication meets the minimum requirements of American National Standard for Information Sciences—Permanence of Paper for Printed Library Materials, ANSI Z39.48–1984 ⊚

CONTENTS

PREFACE / *vii*
ACKNOWLEDGMENTS / *xi*

PART I • THE BACKGROUND TO NEWTON'S SOLUTION

1. A Simplified Solution: The Area Law, the Linear Dynamics Ratio, and the Law of Gravitation / *3*

2. An Overview of Newton's Dynamics: The Problem of the Planets and the *Principia* / *12*

3. Newton's Early Dynamics: On Uniform Circular Motion / *40*

PART II • A GUIDED STUDY TO NEWTON'S SOLUTION

4. The Paradigm Constructed: *On Motion*, Theorems 1, 2, and 3 / *69*

5. The Paradigm Applied: *On Motion*, Problems 1, 2, and 3 / *95*

6. The Paradigm Extended: *On Motion*, Theorem 4 and Problem 4 / *119*

PART III • THE REVISIONS AND EXTENSIONS TO NEWTON'S SOLUTION

7. The *Principia* and Its Relationship to *On Motion*: A Reference Guide for the Reader / *141*

CONTENTS

8. Newton's Unpublished Proposed Revisions:
   Two New Methods Revealed / *166*

9. Newton's Published Recast Revisions:
   Two New Methods Concealed / *182*

10. Newton's Dynamics in Modern Mathematical Dress:
    The Orbital Equation and the Dynamics Ratios / *211*

APPENDIX

An English Translation of Sections 1, 2, and 3 of Book One
from the First (1687) Edition of Newton's
Mathematical Principles of Natural Philosophy / *225*

NOTES / *269*
REFERENCES / *289*
INDEX TO THE GUIDED STUDY AND
THE TRANSLATION / *293*
GENERAL INDEX / *297*

PREFACE

Early in the seventeenth century, the astronomer/mathematician Johannes Kepler demonstrated that the movement of the planet Mars is best described as elliptical motion about the sun located at a focus of the ellipse. Late in the seventeenth century, the challenge still remained for astronomers to determine the nature of the force required to maintain elliptical motion about a focal force center: the Kepler problem. The key to Newton's dynamics resides in his solution of the Kepler problem. It is the goal of this book to make a detailed explanation of that solution available to a wide range of interested students and scholars. Newton's answer provides the analytical basis for the concept of a universal gravitational force. Much has been written on the *ramifications* of this solution, but the details of the solution are rarely made available to any but the expert in the field. The historian of science may be deterred by the mathematical details, the scientist by the conceptual details, and the student by both sets of details. When these details are provided, however, there appears a surprisingly clear and simple analytical structure that frames Newton's speculation concerning the role and nature of force. This structure arises in Newton's early work at Cambridge (pre-1669); it continues to develop after his revival of interest in the problem after 1679; and it achieves its fruition in the first three sections of the first edition of his *Mathematical Principles of Natural Philosophy* (*Principia*) in 1687.

Chapters 1 and 2 of this book set the Kepler problem in historical and conceptual perspective with all reference to mathematical detail postponed. The object is to set forth clearly the challenge of the direct problem of elliptical planetary motion and to supply the conceptual tools employed in its solution, in particular Newton's debt to the works of both

Descartes and Galileo. Chapter 3 presents a detailed discussion of two of Newton's early (pre-1669) analyses of uniform circular motion. In chapters 4 to 6, Newton's solution to the specific direct problem of elliptical planetary motion is examined in detail as it appears in the set of four theorems and four problems that he sent to Halley in 1684 preliminary to the publication of the *Principia*. Chapters 7 to 9 explore the revisions and extensions that are made to these basic elements in the first and revised editions of the *Principia*, and chapter 10 transforms the basic theorems into modern mathematical dress. The book concludes with a translation into English of the first three sections of Book One of the first edition (1687) of the *Principia*. The first edition has rarely been translated and its choice here provides a capstone for the detailed analyses of this book. It also provides a comparison to the existing translations of the third edition (1726) in a direct fashion not available in a variorum edition.

I have used portions of this book in an undergraduate course in the history of science; the students needed only a general high-school background of basic mathematics and science. Specifically, I have used the theorems and problems from the tract *On the Motion of Bodies in Orbit* (*On Motion*) from chapters 4 and 5. In one class period I assigned the details of Theorem 1 (the area law) and the details of Theorem 3 (the force law). In a second class period I assigned the details of Problem 1 (a simple application) and then discussed the solution of Problem 3 (elliptical planetary motion). Other sections were assigned as supplementary reference material. I have also used the entire book as a text for an advanced junior/senior undergraduate course in the history of science, usually as a tutorial. It should serve the same function as a graduate text in departments of history of science.

This book is intended, however, for scholars as well as for students. My choice to study the details of a single problem, however important, may seem overly restrictive. Scholars are interested in the development and growth of Newton's thought on the nature and source of gravitational force, and his reflections on the very nature of scientific analysis itself. A continuity of method, however, lies beneath the changing vocabulary and developing techniques of Newton's work. His method is revealed only by a study of the *details* of the solutions in his early and later work. While that method itself may not be sufficient to reveal Newton's innermost thoughts, nevertheless it provides a measure against which speculations can be held. Consider, for example, the claim often made that Newton's early work reveals a confusion concerning force that was later eliminated, specifically that he attempted to combine two or more different force concepts. It is my contention that an analysis of the details of Newton's solution reveals no such confusion. One must understand the details in order to make an no such confusion. One must understand the details in order to make an informed decision. As a second example, consider the question of the debt owed by Newton to Robert Hooke on the nature of celestial dynamics. The debate revolves about Newton's switch in terminology from *centrifugal* to *centripetal* force following his correspondence with Hooke in 1679. I argue that a close inspection of the details of the post-1679 solution reveals that Newton's method did not change from the method used before 1669. The method survives even if there is a conceptual shift. The truth of that claim lies buried in the trivia of the solution.

I was encouraged to produce such detailed analyses of Newton's solution by my late dear friend and close colleague, Professor Betty Jo Teeter Dobbs. Her death brought a great loss to the world of Newtonian scholarship and to all who knew her. She will be greatly missed. In a letter I received from her after she read the opening sections of my manuscript, she asked if the essentials of the solution could not be presented, as she put it, "without all of that $QR/QT^2 \times SP^2$," by which she meant the analytical details. My reply to her, as it is to all, is that it is not possible and that the task of reading them is really not that formidable. Moreover, the result is worth the effort. Newton's solution to the direct problem of elliptical motion does indeed supply the key to the dynamics that provided the basis for the concept of universal gravitational force.

JB^2

ACKNOWLEDGMENTS

A major portion of this book was written with the support of a grant from the Humanities, Science, and Technology division of the National Endowment for the Humanities (NEH). It is their goal to encourage new translations of and guided studies to major works of science that are of interest to both humanists and scientists. The work of Isaac Newton is widely acclaimed as the most influential in Western science and as such was a logical choice for their support. The project began in January 1990 and the final manuscript was delivered to the publisher in September 1994. During this time, I received help and encouragement from several sources.

In addition to the NEH, I would like to acknowledge the kindness shown by the staff of the library of the Royal Society. They provided a home away from home during my time in London. I would also like to express my gratitude to Lawrence University for a sabbatical leave granted during the project, as well as the summer research support they gave for a student colleague in 1991. Moreover, I had full use of the computer facilities of the university for preparing both the manuscript and the drawings. I am indebted to my colleagues in the department of physics, David Cook and John Brandenberger, for their support and to Bruce Pourciau of the department of mathematics for his patient explanations. In a larger sense, the liberal arts tradition of the university provided the major encouragement. The interdisciplinary nature of the freshman studies program at Lawrence University provided the initial motivation for a physicist to begin to explore with undergraduates the work of Plato, Aristotle, Galileo, and finally Newton. The administration encouraged me to offer a course in the history of science, and my colleagues in the humanities and sciences supported the effort.

It will be clear that this book owes a great debt to Tom Whiteside, the editor of the eight volumes of Newton's mathematical papers. This remarkable publication, a labor of twenty-two years, comprises a collection of Newton's original mathematical papers; Whiteside's translation of them from the Latin provides an invaluable resource for anyone interested in Newton's dynamics or mathematics. Moreover, Whiteside has supplied extensive notes and commentaries that offer the reader historical and mathematical insights. In addition, he has been most generous with his time in correspondence and conversation. I am also grateful to I. Bernard Cohen, who has produced a new English translation of the entire text of the third edition of Newton's *Principia*. I had the privilege of reading portions of an early draft of his translation and benefited greatly from it. He has been a constant source of encouragement. My thanks go to colleagues who have read sections of the first few chapters and provided commentaries—Jo Dobbs, Herman Erlichson, Ivor Grattan-Guiness, and Peter Spargo—and to Michael Nauenberg for his comments on the final chapter. Although they have not seen the entire manuscript and are not responsible for the opinions expressed in it, I have profited from their observations. I also wish to thank Alan Shapiro for the extremely useful commentary he provided as a referee for an earlier version of the manuscript.

This work also drew heavily upon my student colleagues at Lawrence University. I first read the opening sections of the *Principia* with two physics majors, Bob Hanisch and Gene Peterson, when we were at the Lawrence University London Center. Paul Stieg did a senior seminar in which we explored in detail the first three sections of Book One. Outstanding were the contributions of Andrea Murschel, who spent the summer of 1991 as my research assistant at Lawrence University and continued in her senior year to bring her extensive linguistic and analytic skills to bear on the project. In 1992 she was awarded a National Science Foundation fellowship in the history of science and she shows great promise of becoming an outstanding scholar. Credit is also due many of my other students at Lawrence University. In class, they worked through the sections of the tract *On Motion* to be found in chapters 4, 5, and 6, and they served as a check on the accuracy and clarity of my explanations. Their only reward, beyond the joy of learning, was induction at the end of the term into membership of the Loyal Society of S.I.N. (Sir Isaac Newton) with its toast, made with cider of course: "Up with gravity, down with levity."

Finally, and in the place of honor, I want to thank my wife, Dr. Mary Ann Rossi, the classics scholar who translated into English the tract *On Motion* and the selections from the first edition of the *Principia*. Long hours were spent as I, the physicist, argued for what I thought was Newton's intent, and she, the Latinist, argued for what Newton's Latin actually said. In addition, she took time from her other scholarly activities to read and comment upon the various drafts of my manuscript. One is fortunate to have an agreeable companion in one's personal or professional life; I am doubly fortunate to have found one who serves that role in both spheres. This project has been rewarding, but it pales in comparison to our earlier joint venture that produced daughters Lynn and Sandy and sons Rob and Scot. This work is most affectionately dedicated to the memory of our daughter Sandy (12 October 1958–3 February 1995).

PART ONE

The Background to Newton's Solution

Bust of Isaac Newton, by L. F. Roubiliac (c. 1737), currently located in the entrance hall of the Royal Society in London. Newton was elected to the Royal Society in 1662 and served as its president from 1703 until his death in 1727. Copyright © The Royal Society. Reproduced by permission.

ONE

A Simplified Solution

The Area Law, the Linear Dynamics Ratio, and the Law of Gravitation

Isaac Newton's *Philosophiae Naturalis Principia Mathematica* (The mathematical principles of natural philosophy), hereafter referred to as the *Principia*, justifiably occupies a position as one of the most influential works in Western culture, but it is a work more revered than read. Three truths concerning the *Principia* are held to be self-evident: it is the most instrumental, the most difficult, and the least read work in Western science. A young student who passed Newton on the streets of Cambridge is reported to have said, "There goes the man who writ the book that nobody can read." It fits Mark Twain's definition of a classic as a work that everyone wants to have read but that nobody wants to read. The essential core of the *Principia*, however, does not lie beyond the reach of any interested and open-minded individual who is willing to make a reasonable effort.

In 1693, Richard Bentley, a young cleric who was later to become Master of Newton's college, wrote to ask Newton for advice on how to master the work. Newton suggested a short list of background materials, and then, concerning the *Principia* itself, advised Bentley to read only the first three sections in Book One (i.e., the first sixty pages of the four hundred pages that make up the first edition). These sections provide the theoretical background for the astronomical applications that Newton presented in Book Three and regarded as of popular scientific interest. In the introduction to Book Three, Newton repeated the advice that he had given to Bentley:

> I had composed the third book in a popular method so that it might be read by many. But since those who had not sufficiently entered into the principles could not easily discern the strength of the consequences nor put aside long-held prejudices, I chose to rework the substance of that book into the form of propositions in the mathematical way, so that they might be read only by

> those who had first mastered the principles. Nevertheless, I do not want to suggest that anyone should read all of these propositions—which appear there in great number—since they could present too great an obstacle even for readers skilled in mathematics. It would be sufficient for someone to read carefully the definitions, laws of motion, and the first three sections of the first book; then let [the reader] skip to this [third] book.[1]

Newton's sage advice to the general reader to concentrate on the first three sections of Book One of the *Principia* appeared in the first edition of 1687 and remained unchanged in the two revised editions published in 1713 and 1726, all during Newton's lifetime. It is the third and final edition that has been reproduced in many subsequent editions and translated into many other languages. Because this third edition is readily available and because it is seen to represent Newton's most fully developed views, it is almost exclusively taken as a basis for the study of Newton's dynamics. The general reader, however, should not begin with this final edition and its many additions and revisions, but rather with the first edition and its relatively straightforward presentation.

In 1684, Newton sent to London a tract entitled *On the Motion of Bodies in Orbit (On Motion)* that was to serve as the foundation for the first edition of the *Principia* of 1687. This comparatively short tract presents in a clean and uncluttered fashion the basic core of Newton's dynamics and its application to the central problem of elliptical motion. The brief set of definitions that appeared in *On Motion* was expanded in the *Principia* into a much larger set of definitions, laws, and corollaries. Further, the first four theorems and four problems in *On Motion* were expanded into fourteen lemmas and seventeen propositions in the *Principia*. (Theorem 1 of *On Motion* is Proposition 1 of the *Principia* but Problem 4 of *On Motion* is Proposition 17 of the *Principia*). The expanded framework of numbered propositions by itself, however, does not tell the entire story. Even more troublesome for the general reader is Newton's practice of adding new material to the old framework. Having established the expanded set of propositions and lemmas in the early draft of the first edition, Newton elected to hold to that framework as he inserted additional material in his published revised editions. Even in the preface to the first edition, Newton apologized to his readers for such insertions.

> Some things found out after the rest, I chose to insert in places less suitable, rather than to change the number of the propositions as well as the citations. I heartily beg that what I have done here may be read with patience.[2]

After the publication of the first edition, Newton began work on a grand radical revision of the *Principia* in which many of the propositions would have been renumbered and retitled. In contrast to the single method of the first edition, Newton clearly presented three alternate methods of dynamic analysis in this projected revised scheme, each method set forth in a new proposition. Unfortunately, Newton never implemented this new scheme of renumbering the propositions and lemmas in the published revisions. If the challenge of renumbering the propositions and correcting the cross-references was too much in the limited first edition, then it was apparently overwhelming in the expanded revised editions. The new material added to the published revised editions simply was inserted into the old structure of the first edition. The third method of dynamic analysis, so clearly differentiated in the projected revision, was distributed throughout the theorems and problem solutions of the second and third sections of the published revisions. The reader of *On Motion* and, to a lesser extent, of the first edition is not faced with this difficulty. In those works, Newton clearly explicates his analysis with a single method applied uniformly to several problems; until the reader understands his original method and his unpublished restructuring, however, Newton's additions to the much studied revised third edition appear as distractions rather than enrichments.

A SIMPLIFIED SOLUTION

The story of Isaac Newton and the apple is a familiar one. We have all seen the portrayal of an English gentleman who is sitting under a tree and is struck on the head by a falling apple. In a flash, he leaps to his feet and runs off shouting about the theory of universal gravitation. The story has its foundation in Newton's own telling and is attested by a number of memoranda written by those close to him in his later years. The setting is the garden of his country home, the time is 1666, and Newton, a young man of twenty-four, is home after a few years at university. The apple tree that provides his inspiration stands in his front garden, and the fruit it bears is a yellow-green cooking apple called the Flower of Kent. One version of the story, told by Newton in his later years and recorded by an associate, John Conduitt, includes the following statement:

> Whilst he was musing in a garden it came into his thought that the power of gravity (which brought an apple from the tree to the ground) was not limited to a certain distance from the earth but that this power must extend much farther than was usually thought. Why not as high as the moon said he to himself and if so that must influence her motion and perhaps retain her in her orbit, where upon he fell to calculating what would be the effect of that supposition.[3]

There is evidence that Newton made a calculation comparing the moon's centrifugal force, a celestial event, with the local force of gravity, a terrestrial event. Since it is a calculation that could have been inspired by any falling object, why not an apple? That early calculation of 1666 did not supply the mathematical basis for the general demonstration that the force necessary to maintain a planet in an elliptical orbit about the sun located at a focus of the ellipse is inversely proportional to the square of the distance between the sun and the planet (i.e., the law of universal gravitation). It was late in 1684, after Edmund Halley's famous visit to Newton's rooms at Cambridge University, before Newton gave anyone a copy of such a proof—a proof which Newton claimed to have produced in 1679. The inspiration of the falling apple of 1666 required more than a decade to reach its final goal.

In 1684, Newton sent Halley a solution to the problem of planetary motion in the tract *On Motion*. That solution is expressed neither in the mathematics of classical geometry nor in the mathematics of contemporary differential and integral calculus. As such it is a challenge to the modern physicist as well as to the classical scholar. The outline of the solution, however, is not complicated; it is the details that provide the challenge. Newton adapted the linear kinematics of Galileo to the inertial dynamics of Descartes and determined the nature of the force necessary to maintain planetary motion as described by Kepler. If a constant linear acceleration A is acting on a body of mass m, then its displacement D is proportional to the constant acceleration A and the square of the time t (i.e., $D = (\frac{1}{2})At^2$). If one adds to this simple kinematic relationship the dynamic relationship that the acceleration A is proportional to the force F (i.e., $F = mA$), then the force F is directly proportional to the displacement D and inversely proportional to the square of the time t (i.e., $F = (2m) D/t^2$). Newton's genius manifests itself in adapting this simple proportional relationship of constant rectilinear force, distance, and time to the more complex problem of the nature of the planetary force, which is not constant. Newton's unique contribution was the assumption that the variable force could be considered to be approximately constant over a very short period of time. The three elements that Newton generated to produce the solution can be set forth quite simply: first, the relationship that expresses the time in terms of the area; second, the relationship that expresses the force in terms of the displacement and the time (and hence in terms of the area); and finally, the relationship that expresses the force necessary for planetary motion in terms of distance (i.e., the demonstration that the gravitational force is inversely proportional to the square of the distance).

Theorem 1

The first element is the law of equal areas in equal times, demonstrated in figure 1.1. If the force acting on a body is always directed to a fixed point S, then the time required to travel from point P to point Q is proportional to the shaded area SPQ. If successive areas are generated in equal

A SIMPLIFIED SOLUTION

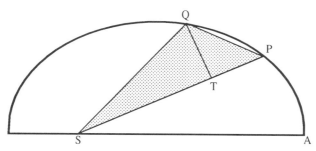

Figure 1.1 If a body moves from point P to point Q under a centripetal force directed toward the fixed point S, then the shaded area SPQ is proportional to the time.

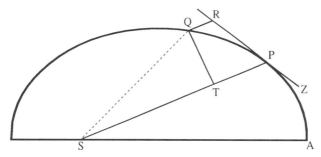

Figure 1.2 The force F required to maintain any orbit APQ about a center of force S is proportional to the displacement QR and inversely proportional to the area SQP.

times, then the areas swept out by the line from the body to the center of force are equal. This relationship was first recognized by the astronomer Johannes Kepler in 1609, but it was not until after 1679 that Newton demonstrated its general application to any motion under any force directed toward a fixed center. The area law is the link that was missing in Newton's earlier analysis of motion and it is the key element in his celestial dynamics; it appears as Theorem 1 in the 1684 tract *On Motion* and as Proposition 1 in the 1687 *Principia* (see chapter 4 for details).

Theorem 3

The second element is the basic relationship that I have elected to call the "linear dynamics ratio." Figure 1.2 is similar to Newton's diagram for Theorem 3 in *On Motion* and for Proposition 6 in the *Principia*. The line RPZ is the tangent to the curve APQ. If no force acted on a moving body at point P, then the body would continue in a given time interval along the tangent line to the point R. Because a force does act continuously on the body, however, it moves instead to the point Q. The displacement QR represents the deviation of the body from the tangential path PR due to the action of the force. Galileo had demonstrated in his experiments with inclined planes that for motion under a given constant force, the distance traveled is proportional to the square of the time. Newton assumes that as the point Q shrinks back to the point P, then the force can be treated as if it were constant. Thus, the distance QR is proportional to the square of the time and to the magnitude of the force at point P, or what is equivalent, the force is directly proportional to the distance QR and inversely proportional to the square of the time. From Theorem 1, the time is proportional to the triangular area SPQ and thus can be expressed in terms of the altitude QT and the base SP. The result is that the force F at point P can be expressed as follows:

$$\text{Force} \propto \text{distance} / (\text{time squared}) \propto QR / (QT^2 \times SP^2)$$

The challenge is to express the ratio QR/QT^2 in terms of SP and constants of the orbital figure, and hence to express the linear dynamics ratio $QR/(QT^2 \times SP^2)$, and thus the force, in terms of the radial distance SP (see chapter 2 for a review and chapter 4 for a detailed discussion of this theorem).

Problem 3

The third element is a demonstration by Newton of a relationship between portions of an ellipse. Figure 1.3 is a drawing of a planetary ellipse APQ with a focus at point S. The line LSL drawn through the focus S and perpendicular to the major diameter of the ellipse is called the *latus rectum L*. Newton demonstrates in Problem 3 and in Proposition 11 that as the point Q shrinks back to the point P, the ratio QR/QT^2 becomes equal to the reciprocal of the *latus rectum L*, which is a constant for a given ellipse. Thus, the force can be obtained quite simply from the linear dynamics ratio above:

$$\text{Force} \propto QR / (QT^2 \times SP^2) = 1 / (L \times SP^2) \propto 1 / SP^2$$

This result states that the force required to maintain a planet in an elliptical orbit about the sun located at a focus of the ellipse is proportional to the inverse square of the distance between the planet P and the sun S. Thus is demonstrated the mathematical basis for the law of universal gravitation, the essence of celestial interactions, which Newton provides for future astronomers and physicists (see chapter 2 for a review and chapter 5 for a detailed discussion of this problem).

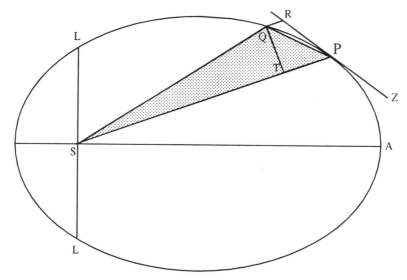

Figure 1.3 As point Q shrinks back to point P, the ratio QT^2/QR becomes equal to the line LSL, which is a constant (the *latus rectum*) for a given ellipse.

The details of the demonstrations of the relationships above are more demanding than is evident in this verbal gloss. Taken step by step, however, the analysis will become clear to the reader. At times Newton makes analytical leaps that for him are obvious and it is then my duty to supply the intervening steps. Thus, it is the number of steps rather than the size of any single step that offers the challenge. The reward for the patient reader is an insight into the solution of the problem of planetary motion, a problem that challenged astronomers for millennia. That solution is now universally held to have provided a major turning point in astronomy and natural philosophy in the late seventeenth century.

THE RECEPTION

Professional scholars, however, did not greet the publication of the 1687 *Principia* with unreserved praise. The dominant figure in seventeenth-century natural philosophy was the French scholar René Descartes, whose mechanical description of planets carried in a swirling vortex of celestial ether provided the model for many other natural philosophers. Two other outstanding figures in European mathematics and natural philosophy at the time of the publication of the first edition of the *Principia* were the Dutch scholar Christiaan Huygens and the German scholar Wilhelm Gottfried Leibniz. Both felt that Newton's description of the mathematical nature of gravitational force had failed to address the fundamental question of the physical cause of the force. It would appear that Huygens accepted the inverse-square law as a genuine discovery, although he believed that its cause remained to be investigated. Leibniz initially praised the 1687 *Principia* as one of the most important works of its kind since Descartes. He criticized Newton, however, for his rejection of Cartesian vortices and for his failure to provide an alternate physical cause for the gravitational attraction. In England, the astronomer and mathematician Edmund Halley served as the editor of the 1687 *Principia* and it was published under the imprimatur of the Royal Society. Even with this auspicious beginning, it was not without controversy that the *Principia* was finally published. The English scientist Robert Hooke claimed priority for the discovery of the inverse square nature of the gravitational force, a claim that Newton vehemently rejected. Despite individual reservations, the overall reception by the scholarly community was positive, and Newton established himself as one of the leading mathematicians of Britain and Europe. As one modern scholar of Leibniz's work put it, "Already in . . . 1695, Leibniz had abandoned the project of presenting a theory capable of competing with Newton's. Despite his subtle philosophical and theological objections, in the eighteenth century Leibniz had left Newton master of celestial mechanics."[4]

As the scholarly reputation of the *Principia* grew, even those who professed little or no mathematical ability came to pay homage. The English philosopher John Locke, in exile in Holland at the time of the book's publication, obtained assurance from Huygens that the mathematical propositions of the *Principia* were valid and then applied himself to understanding Newton's conclusions. Locke eventually referred to "the incomparable Mr. Newton" in the preface to his *An Essay Concerning Human Understanding*. The French writer and philosopher François Voltaire waxed even more eloquent when he drew the following comparison between Newton and the German astronomer Johannes Kepler: "Before Kepler, all men were blind. Kepler had one eye, Newton had two." The English poet Alexander Pope's often quoted heroic couplet, published shortly after Newton's death, revealed even more forcefully the popular view that Newton and the *Principia* opened doors that had long been closed: "Nature, and Nature's Laws lay hid in night: / God said, Let Newton be! and all was light." In the dedicatory poem to the first edition of the *Principia*, Edmund Halley reflected on Newton's "unlocking the treasury of hidden truth" and concluded that "nearer to the Gods no mortal may approach."

As the eighteenth century drew to a close, not everyone continued to praise the new world that appeared in Newton's work. Figure 1.4 is an early nineteenth-century caricature of Newton by the philosopher-poet-artist William Blake, who, putting imagination above reason, reacted negatively to the eighteenth-century veneration of Newton. In the portrait,

Figure 1.4 William Blake's portrait of Isaac Newton. Courtesy Tate Gallery, London.

triangles abound as the symbol of the geometrical and mathematical mentality that Blake opposed. Newton holds a triangular compass as he draws a triangular figure on the parchment, his fingers make triangles with the object in his hand, his legs form triangles with each other and with the rock on which he sits, the muscles of his body take on geometric forms that defy anatomical description, and triangular eyes scheme as they look down a triangular nose at geometric plans that triangular hands create below. For Blake, Newton symbolized the eighteenth-century regard for human reason that placed God above and separate from women and men, while Blake regarded human imagination as the essential divine quality by which God was made manifest.[5]

Neither Newton nor his *Principia* deserves the extreme judgments of Pope and Blake; the work is ranked as one of the major intellectual achievements of Western culture. The Enlightenment of the eighteenth century and the Romanticism of the nineteenth century both have their roots in the acceptance or rejection of the new worldview that paid homage to Newton's scientific writings. Just as his *Optics* provided a model for the experimental method, so his *Principia* laid the foundations for the theoretical method. It was, in fact, the wave of the future.

TWO

An Overview of Newton's Dynamics
The Problem of the Planets and the *Principia*

The authority of the *Principia* stemmed initially from its ability to provide a solution to one of the major challenges that astronomers faced, the problem of the planets. The vast majority of stars maintain a fixed position relative to one another. The planets, however, move against the background of the constellations (the word "planet" comes from the Greek word meaning "wanderer"). Their appearances, retrograde loops, and disappearances have provided a continuous challenge to astronomers for over three thousand years. An early version of the problem of the planets was set for astronomers by Plato, who challenged them to find the set of uniform circular motions that would "save the phenomena" of the apparent wandering motions of the planets. It is in this tradition that the work of the astronomer Ptolemy was done in the second century, and it was in this tradition that the work of Copernicus was done in the middle of the sixteenth century, even as he proposed a shift from a fixed earth to a fixed sun. Kepler was also working in this tradition early in the seventeenth century, even as he described the noncircular elliptical motion of the planets in his *New Astronomy* of 1609. Despite his description of planetary motion as elliptical, the circle remained the primary element for Kepler in his understanding of the archetypal cause central to God's plan of the universe, and that understanding was strengthened by his complementary concern for physical cause.[1]

The second half of the seventeenth century was dominated by a quite different approach to astronomical cause, a mechanical approach concerned with celestial subtle fluids and vortices. In his *Philosophical Principles* of 1651, the French mathematician and philosopher René Descartes postulated three types of matter that constitute the heavens: the sun and the fixed stars formed from the first type, the planets and comets formed from the third type, and the rest of the heavens filled with the second type. The motion of the planets, caught in a swirling vortex of the second type of matter, was compared to the motion of straws caught in a whirlpool of water. Descartes's accounts of celestial motion were qualitative, and subsequent attempts by astronomers to derive quantitative results from them failed. Nevertheless, Cartesian mechanical cause, in contrast to the Keplerian archetypal cause that preceded it, provided the basis for much of the natural philosophy of the second half of the seventeenth century. It was in this Cartesian tradition that Newton began his study of mathematics, mechanics, and astronomy.

Shortly after entering Cambridge University in 1661, Newton's introduction to higher mathematics came from Descartes's *Geometry* and his introduction to mechanics came from Descartes's *Philosophical Principles*, which also introduced him to a world of celestial vortices. For the next fifteen years his thinking on astronomy was colored and structured in part by Cartesian images.[2] Following his solution of the planetary problem in 1679, however, Newton broke with Cartesian tradition. For Newton, the problem of the planets was neither the Platonic challenge to seek celestial circularity nor the Cartesian charge to find mechanical causes. Rather, it was the dynamic challenge of seeking the mathematical description of the force that produces the Keplerian planetary ellipses. In his published work, he set aside his concern with physical cause and emphasized the mathematical form of the force. Newton offered this limited type of mathematical answer amid the seventeenth-century concern for the physical causes of gravity, and his approach was not immediately or universally found to be fulfilling. The technique that Newton employed, a unique combination of limits and approximations, was rapidly replaced on the continent by alternate techniques that employed the calculus more directly. Nevertheless, the *Principia* set the stage for the mathematical, mechanical model of the world that would follow. Newton outlined the challenge quite clearly in the preface to the first edition.

> I offer this work as mathematical principles of philosophy. For all the burden of philosophy seems to consist in this—from the phenomena of motions to investigate the forces of nature, and then from these forces to demonstrate the other phenomena.

Thus, Newton set two general challenges for his work: first, to determine the details of the unknown forces of nature directly from knowledge of observed physical phenomena, and second, to determine the details of additional physical phenomena not yet observed from knowledge of these revealed forces. The specific problem that claimed his attention, of course, was the problem of the planets. Given the knowledge from Kepler that a planet moves in an elliptical path subject to a force provided by the sun (located at a focus of the ellipse), Newton first determined that the centripetal force must be inversely proportional to the square of the distance. Then, armed with that functional form of the law of universal gravitation ($1/r^2$), he was able to turn his attention to the application of the force to other physical phenomena.

Newton built his dynamics upon three earlier studies of motion: uniform circular motion, uniformly accelerated rectilinear motion, and the motion of colliding bodies. Uniform circular motion had its earliest application in Greek and scholastic astronomy. Aristotle separated the terrestrial world of chaotic motion, which was below the lunar sphere, from the celestial world of uniform motion, which was above the lunar sphere. Aristotle believed that the hallmark of motion in the celestial world was the circle; circular motion, having no beginning and no end, was eternal. The stars were observed to move about the earth in a combination of a primary daily circular motion and a secondary annual circular motion. The sun, the moon, and the planets appeared to move in a more complicated fashion, but astronomers were able to resolve these motions into component motions composed of additional uniform circular movements called epicycles. Copernicus still held the belief in celestial circularity when he proposed the drastic revision of celestial mechanics that transferred the fixed center of motion from the earth to the sun. Indeed, his work is full of uniform circular motions in the form of primary and secondary epicycles that combine to predict the motion of the planets. When Newton employed uniform circular motion in his dynamics, however, it was in a way dramatically different from any earlier application.

Uniformly accelerated rectilinear motion, in contrast to uniform circular motion, manifests itself in terrestrial motion. The motion of a body falling freely near the surface of the earth is a good approximation to motion in a straight line under the influence of a constant force. This topic was explored by many scholars, including Aristotle, Galileo, and Newton. In his *Physics*, Aristotle examined the motion of bodies of different weights moving in a resistive medium with constant speeds. Falling bodies display such motion after achieving terminal speeds during descent in a column of water or glycerine. The terminal speed of a heavier body is proportional to its weight, and thus the rate of fall of a heavier body is greater than that of a lighter body.[3] Aristotle did not discuss this particular physical example, but he arrived at the same conclusion by investigating the nature of resistive motion in general. In *The Two New Sciences*, Galileo addressed the more general problem of the motion of bodies in the absence of resistance, in which bodies of different weights move with constant acceleration. Falling bodies display such motion in a column of extremely rarified air in an evacuated chamber. Galileo offered both theoretical and experimental evidence that the rate of the fall of bodies in such a void was independent of their weights. Furthermore, Galileo demonstrated in a series of experiments with inclined planes that the distance traveled is directly proportional to the constant acceleration and to the square of the time. Newton employed this functional form of Galileo's description of uniformly accelerated linear motion in his own solution to the more complicated problem of elliptical planetary motion.

The third type of motion involved the study of collisions. On this topic, Newton was strongly influenced by Descartes's work on the motion of colliding bodies. Descartes set forth the principle of linear inertia: A body continues in a natural state of uniform rectilinear motion unless acted upon by an external influence. The momentary clash of two bodies provides the source of the external influence that changes their natural motion. The dynamics of the interaction last only during the brief period of the contact and are common to everyday experience. Descartes attempted to extend his principles of terrestrial collisions to planetary dynamics by postulating a celestial plenum whose vortex motion carried the planets in its swirl by direct contact. Newton ultimately rejected Descartes's celestial vortices and many of the particular solutions produced by Descartes on the phenomena of colliding terrestrial bodies. He adopted Descartes's principle of linear inertia, however, and applied it with Galileo's kinematics of rectilinear motion to the problem of planetary motion. Newton's genius manifested itself in his application of these simpler types of motion to the more complex elliptical motion of the planets, a motion which is not rectilinear, circular, or caused by collisions.

THE DIRECT PROBLEM

In December of 1684, Newton sent Edmund Halley a tract containing a set of theorems and problem solutions; this tract was the basis for the opening sections of the 1687 edition of the *Principia*. The fundamental type of problem addressed in the tract *On Motion* and in the first three sections of Book One of the *Principia* may be stated as follows: Given a path of a particle and a center of force relative to that path, find the dependence of the force upon the distance between the particle and the center of force. The solutions to such problems proceed directly from the observed path to the unseen force, and therefore the problems were called "direct" by the early investigators. Conversely, solutions to problems that proceed in the inverse direction using descriptions of the force function to find the path about the force center were called the "inverse" problems.

The direct problem that challenged mathematicians in the seventeenth century was determining the nature of the gravitational force acting on planets. In 1601, Kepler obtained a detailed set of observations of the motion of the planet Mars from the Danish astronomer Tycho Brahe. From his analysis of that data Kepler determined that the path of Mars is an ellipse, with the sun located at a focal point. He published the result in 1609. The direct problem, however, remained unsolved until after 1679, when Newton determined the functional dependence on distance of the force required to sustain such an elliptical path of Mars about the sun as a center of force located at a focal point of the ellipse. Building on Newton's description of the nature and universality of the gravitational force, scientists of the eighteenth century shifted their interest almost exclusively from direct to inverse problems. They used the combined gravitational forces of the sun and the other planets to predict and explain perturbations in the conic paths of planets and comets. That interest continued through the nineteenth and twentieth centuries, and today scientists still concentrate upon the inverse problem rather than the direct one. In fact, the early terminology is often reversed by modern scientists; what was then called the inverse problem is now called the direct problem because it is seen as a direct application of the law of universal gravitation to a particular physical problem.[4] For Newton, however, the challenge of finding the nature of the force functions remained the primary and direct problem.

Kepler had reported in 1609 that the planet Mars moves in an ellipse, with the sun located at a focus, and that the radius vector from the sun to the planet sweeps out equal areas in equal times. The nature of the force function required for uniform circular motion had been determined by Huygens in 1659 and also, independently, by Newton in 1669.[5] No one before Newton, however, had demonstrated the specific mathematical nature of the force function required for elliptical motion. In the preliminary 1684 tract *On Motion* and in the 1687 *Principia*, Newton employed a method of polygonal approximations to demonstrate that Kepler's law of equal areas holds for any force directed toward a fixed center. He then used that result and another set of approximations to extend his dynamics into a general method for determining the nature of the force required to maintain a specific type of orbital motion about a given center of force. Newton used this general method to solve a number of direct problems, the capstone of which was the direct Kepler problem. Among the preliminary examples, chosen by Newton for their mathematical utility rather than any physical significance, were the following:

Orbit	*Force Center*	*Force Function*
Circle	The Center of the Circle	A Constant (the speed squared/the distance)
Circle	Any Point on the Circumference	The Inverse Fifth Power of the Distance

Orbit	Force Center	Force Function
Spiral	The Pole of the Spiral	The Inverse Third Power of the Distance
Ellipse	The Center of the Ellipse	Directly as the Distance

Having demonstrated his general method by these examples, Newton proceeded to present his solution to the distinguished Kepler problem of planetary elliptical motion with the center of force (the sun) located at a focus of the ellipse. He demonstrated that the gravitational force depended upon the inverse square of the distance between the planet and the sun.

NEWTON'S DEBT TO DESCARTES

A survey of Newton's student texts illustrates the early influence that Descartes had upon Newton's analysis of motion. There are three specific ideas that readily demonstrate that influence in his early work: uniform rectilinear motion, change in motion, and uniform circular motion. Newton continued to employ the first two ideas in his mature work, but the third idea, in particular with regard to "outward endeavor," was dramatically revised. An analysis of the details of Newton's calculations reveals his view of the nature of the force required for circular motion more clearly than does the terminology he adapts from Descartes to describe his calculations.

Item 1. Uniform Rectilinear Motion. For Descartes the natural state of motion of a body is to remain at rest or, if set initially into motion by an external cause that is then removed, to remain in uniform rectilinear motion. Thus, an object of and by itself will not move in a curved path unless it is constantly acted upon by an external cause. He sets forth this principle in his *Principles of Philosophy* as follows:

> If it [a body] is at rest we do not believe that it will ever begin to move unless driven to do so by some external cause. Nor, if it is moving, is there any significant reason to think that it will ever cease to move of its own accord and without some other thing which impedes it.[6]

This basic principle of linear inertia appears in the first edition of Newton's *Principia* as follows:

> *Every body perseveres in its state of resting or of moving uniformly straight ahead except insofar as it is compelled by impressed forces to change that state.*

It is interesting to note that both statements appear to have been anticipated by Aristotle, who in his *Physics* makes the following statement: "Hence, a body would either continue in its state of rest or would necessarily continue in its motion indefinitely, unless interfered with by a stronger force."[7] Aristotle, however, is arguing that a void cannot exist, for if it did then the above state of rest or uniform motion would occur. Since such states are not observed in nature, Aristotle claims that a void cannot exist.

Item 2. Change in Motion. Newton's early view of the mechanism by which the external influence changes a body's state of rest or motion also stemmed from Descartes's discussion of collisions. Descartes considered force in terms of the collision of two bodies. Central to his analysis of such collisions is his statement that "no more action is required to produce movement than to bring about its cessation."[8] This statement is echoed in Newton's very early dynamical writings, in his *Waste Book,* as "there is exactly required so much and no more force to reduce a body to rest as there was to put it upon motion."[9] That statement is then expanded into the following generalization:

> So much force as is required to destroy any quantity of motion in a body so much is required to generate it; and so much as is required to generate it so much is also required to destroy it.[10]

The word "force," as Newton uses it in the statement, is the change in the quantity of motion in a body. For a modern physicist, it is the "impulse" that produces the change in the quantity of motion in a body, where the quantity of motion is rendered into modern terms as linear momentum (i.e., the product of the body's inertial mass and its velocity). A modern physicist defines impulse as the product of the force and the time that the force acts on the body, and defines force as "time rate of change" of the linear momentum. A potential source of confusion for the modern reader resides in this difference in terminology between Newton's use of force and the modern distinction between force and impulse.[11]

Descartes also describes as "resistance" the tendency of a body to maintain its state of motion or of rest, a description consistent with the principle of inertia but also suggestive of force. Thus, in a collision between a body at rest and a body in motion, the interaction is described in terms of the "force of resistance" of the first body to remain at rest and the "force of motion" of the second body to remain in motion. The use of the word "force" in that context is also at variance with the word's modern usage. Newton adopted similar terminology to discuss the collision of two bodies. On the one hand, he spoke of an elastic force produced by the deformation of the bodies during the collision; in that sense his use of the term "force" is similar to modern usage. On the other hand, he also spoke of the Cartesian "force of a body's motion," and in that sense his use of the term "force" is at variance with modern usage.

Item 3. Uniform Circular Motion. Even when a body is moved along a curved path, there is a sense in which the natural rectilinear motion perseveres. For Descartes, the "determination" of the motion at an instant is directed in a rectilinear fashion along the tangent to the curve. In support of such a view, Descartes discussed the motion of a ball whirled in a circle by a sling. When the sling is released the ball flies off along a line tangent to the original circular path. The ball deviates from the rectilinear tangential path only because the sling constrains it to do so.

In addition to discussing the natural tendency to move rectilinearly along a tangent of the circle, Descartes also discussed a natural tendency to move rectilinearly along a radius of the circle. In support of this view, Descartes discussed the motion of a ball in a rotating tube.[12] As the tube rotates in a horizontal plane about a vertical axis, the ball accelerates out along the radius of the tube. Descartes saw this increase in radial motion of the ball along the tube as an indication that the ball experienced an outward tendency as well as a tangential tendency. In the example of the ball whirled in a sling, the ball was constrained by the sling. Hence, the outward tendency remained potential. In the rotating tube no such restraint existed and the ball could move outward. Hence, the outward tendency was active. The Dutch physicist and mathematician Christiaan Huygens gave the name "centrifugal force" to this outward-seeking tendency or endeavor.

In modern analysis, however, "centrifugal force" is often called "fictitious force" because the motion is only a manifestation of a rotating coordinate system. When the modern physicist analyzes such motion in an inertial coordinate system outside of the rotating tube, then there is no radial force. The only force on the ball is that exerted by the side of the tube in a direction perpendicular to the radius of the tube. It is a challenge, however, for the instructor in an introductory physics class to convince beginning students that it is "obvious" that there is no outward radial force acting on the ball. Professor William Fogg Osgood in his mechanics textbook of 1937, venting his frustration with students who repeatedly failed to appreciate this point, sets out the modern perspective in a rather dramatic fashion.

> And now, after all is said and done, comes the [students's] rejoinder: ["But the ball did go out of the tube."] There is no answer to these people. Some of them are good citizens. They vote the ticket of the party that is responsible for the prosperity of the country; they belong to the only true church; they subscribe to the Red Cross drive—but they have no place in the Temple of Science; they profane it.[13]

In discussing early methods of dynamic analysis, however, as one modern scholar points out, it is misleading to approach the seventeenth-century view of centrifugal force from the modern perspective of rotating coordinate systems.

> Continental mathematicians, notably Descartes . . . and Huygens shared common ideas about curvilinear motion. Despite some differences on specific points, they believed that curvilinear motion results from the interplay of a tendency towards a center and an outwards tendency due to the rotation of the body and to its rectilinear inertia. . . . Descartes, Huygens, and their contemporaries did not consider centrifugal force as dependent on the choice of a rotating reference frame, as is commonly done in more modern formulations of mechanics.[14]

For Descartes the inward and outward tendencies of a celestial body rotating in a celestial vortex depends upon the quantity of matter of the body, in conjunction with its volume and surface.[15] For Huygens the force of gravity acting on a body originated from the centrifugal force of a rotation fluid surrounding the body and was equal to the difference between the centrifugal force of the fluid and that of the body.[16] Newton's early work on uniform circular motion was expressed in terms of Descartes's "outward endeavor." It was not until Newton began the preliminary work on the *Principia* that he created the term "centripetal force" to stand in contrast to Huygens's term "centrifugal force." Much has been made of this change in language following Newton's correspondence with Robert Hooke in 1679.[17] In fact, Westfall claims that Hooke taught Newton a considerable lesson in dynamic analysis.

> Hooke said nothing about centrifugal force. Orbital motion results from the continual deflection of a body from its tangential path by a force toward some center. Newton's papers reveal no similar understanding of circular motion before this letter. Every time he had considered it, he had spoken of a tendency to recede from the center, what Huygens called centrifugal force; and like others who spoke in such terms, he [Newton] had looked upon circular motion as a state of equilibrium, between two equal and opposing forces, one away from the center and one toward it. Hooke's statement treated circular motion as a disequilibrium in which an unbalanced force deflects a body that would otherwise continue in a straight line. *It was not an inconsiderable lesson for Newton to learn.* [emphasis added][18]

There is a sense in which the preceding statement is correct, but there also is a sense in which it is misleading. The Cartesian language that Newton employs in his early analysis of uniform circular motion clearly indicates that he attributes to the rotating body "a tendency to recede from the center." Yet when he analyzes the problem, his technique is identical to that which he employs in his more mature work: a tangential displacement is combined with a radial displacement, and the nature of the force is revealed. In the early work, the radial displacement may be a potential outward one that is not realized because of the inward force; whereas in the later work, the radial displacement may be an actual inward one that is realized. Nevertheless, the mathematical and dynamical elements employed

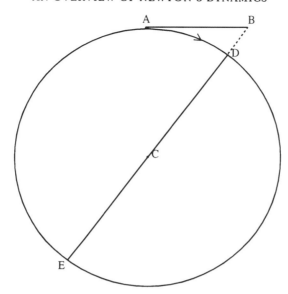

Figure 2.1 A body moves in uniform circular motion about a center C. The radial element DB measures the deviation of the circular arc AD from the tangent AB.

in the analysis are the same in both cases. If the "not inconsiderable lesson for Newton to learn" was the distinction between the two points of view, then the statement is correct. If the argument is that Newton was led to the actual details of the analysis by Hooke, then it is misleading: Newton demonstrates very early that he does understand how to solve direct problems.

Consider, for example, the analysis of uniform circular motion that Newton produced sometime before 1669. Figure 2.1 is based on his diagram from that early paper. Newton sets forth the problem as follows:

> *The endeavor of body* A . . . *from the center* [C] *is as great as would carry it away from the circumference to the distance* DB *in the time* AD . . . *inasmuch as it would reach that distance in that time if only it could move freely in the tangent* AB *with no impediment to the endeavor.*[19]

Two possible readings can be made of this statement. The first identifies the "endeavor from the center" as that which would produce the radial displacement from point D to point B along the extension of the diameter of the circle. In that reading the "impediment to the endeavor" prohibits the outward radial displacement DB and the body travels along the circular arc AD. An alternate reading identifies the "endeavor from the center" as that which produces the linear displacement from point A to point B along the tangent. In that reading the "impediment to the endeavor" prohibits the tangential displacement AB and again the body travels along the circular arc *AD*. In both cases it is the displacement *DB* that measures the "impediment," be it an outward centrifugal force or an inward centripetal force.

From the text of this early work, it is clear that Newton is employing the first of these two readings, the one expressed by Descartes. From the text of his later work, it is clear that Newton is employing the second reading, the one expressed by Hooke. In both cases, however, Newton identifies the displacement *DB* as directly proportional to the force and inversely proportional to the square of the time. The resultant nature of the force is the same in both the early and later solutions. It is proportional to the diameter of the circle and inversely proportional to the square of the period of uniform circular motion. The lesson Newton has learned is an important one, but it is one of perspective and not of analysis.[20]

In his mature work, Newton set aside the Cartesian view of an outward endeavor, but he continued to use Cartesian uniform rectilinear motion and the Cartesian view that a force is required to produce a change in rectilinear motion. He did owe an initial and lasting debt to Descartes, and his failure to acknowledge that debt in the *Principia* is unpardonable. It is true that Newton carried the analysis of orbital motion to a level never imagined by his predecessors, but he acknowledged in the *Principia* a debt to the work of Galileo, who contributed far less to his thought than Descartes. It has long been argued that "Newton eventually came to detest Descartes both as a physicist and a philosopher."[21] Newton's implicit criticism of Cartesian philosophy is distributed throughout the *Principia*. As one scholar has expressed it, Newton's criticism of Descartes is "not absent but only hidden."[22] A clear point of refutation of the Cartesian celestial system manifests itself in the *Principia* in Newton's investigation of the nature of motion in vortices. The Cartesian celestial system is based on the presumption that it is the swirling vortex motion of an all-pervasive celestial ether that collides mechanically with celestial bodies and thus carries the planets around the sun and the planetary moons around the planets. In Proposition 52 of Book Two, Newton states that he has "endeavored in this Proposition to investigate the properties of vortices, that I might find whether the celestial phenomena can be explained by them."[23] He finds that vortices cannot explain the established relationship between the period and the radius of planetary satellites (the so-called 3/2 power law of Kepler). Thus, in Proposition 53 of Book Two he concludes that "it is manifest, that the Planets are not carried round in corporeal vortices."[24] But never does he mention Descartes by name, although it is clear that the Cartesian system is under attack.

Before 1679, Newton had accepted Descartes's theory of swirling dense celestial ether in collision with celestial bodies as the mechanical cause of gravitational interaction. Following that date, Newton no longer perceived the heavens as a mechanical vortex, nor did he continue to employ the Cartesian terminology. It has been argued that Newton's dramatic rejection of the Cartesian explanation for gravitational interaction came as a result of Newton's success in deriving an exact solution to the problem of planetary motion. Specifically, the argument is that the key to Newton's transition in thought lies in his demonstration of the exact nature of Kepler's law of equal areas.[25]

In 1684, Newton sent Edmund Halley the tract *On Motion*, the forerunner of the first edition of the *Principia*. This preliminary tract contained four theorems and seven problem solutions. The four theorems and the first four solutions were related to celestial problems of planetary motion with the assumed absence of resistance, and the final three problems were related to the terrestrial problem of projectile motion in air with the assumed presence of resistance. Measurements made of the motion of projectiles in the terrestrial atmosphere deviated from the idealized parabolic paths postulated under the assumption of no resistance because the atmospheric air did provide resistance to motion.[26] Yet the motion of celestial bodies did not deviate from the idealized elliptical paths predicted under similar assumptions of no resistance due to the celestial ether. This observation posed a problem. If Newton accounted for this celestial behavior by assuming that the ether was so diffuse that it caused no resistance, then he could no longer assume that the ether was dense enough to provide the mechanical collisions needed for the gravitational interaction.

Such a conclusion would have called for a major revision in the way Newton saw the celestial world. It was not a step to be taken lightly. Nevertheless, Professor Jo Dobbs has recently argued that Newton did take such a step and that the exact nature of the area law played a major role in his decision to reject mechanical celestial collisions.

> Newton made a dramatic break with the orthodox mechanical philosophy of his day, the philosophy that was generally understood among advanced thinkers at the time to be the most promising method of approaching the study of the natural world. He did not reject the entire system of mechanical thought, but he did reject one of its most basic assumptions: that force could be transferred only by the impact of one material body with another.[27]

Within the *Principia*, however, Newton did not offer a mechanism to replace the Cartesian mechanical gravitational interaction, being content (at least there) simply to set out the mathematical dependence of gravitational force on the distance (i.e., it is inversely proportional to the square of the distance).[28] Thus, Newton's reaction to being led astray for two decades by Descartes's postulated vortices of dense celestial ether may have played a major role in his refusal to accord Descartes any credit whatsoever.

Newton's demonstration of Kepler's area law had other far-reaching effects on his orbital mechanics, and it is this demonstration that accounts for the progress he made following his correspondence with Hooke in 1679. As Whiteside, the editor of Newton's mathematical papers, states, "Hooke himself was the immediate *catalyst* in exciting the fundamental change through which Newton's astronomical thought went in the winter of 1679/ 1680."[29] It was Hooke's challenge, to which Newton never directly replied in his correspondence with Hooke, that motivated the derivation of the area law. Whiteside expresses the challenge as follows:

> The problem had been squarely, unambiguously put to him [by Hooke]: Does the central force which, directed to a focus, deflects a body uniformly traveling in a straight line into an elliptical path, vary as the inverse-square of its instantaneous distance from that focus? To repeat Newton's already quoted words, '*I found now that whatsoever was the law of the forces which kept the Planets in their Orbs, the areas described by a Radius drawn from them to the Sun would be proportional to the times in which they were described.* And . . . that their Orbs would be such Ellipses as Kepler had described [when] the forces which kept them in their Orbs about the sun were as the squares of their . . . distances from the sun reciprocally. . . . [emphasis added][30]

It is the law of equal areas that provides the mechanism for extending the orbital elements in Newton's early analysis of uniform circular motion to the more general problem of nonuniform elliptical motion. In his early analysis, time was measured by equal *angles* in equal times; in his later analysis, time is measured by equal *areas* in equal times.[31] Whiteside concludes, "At long last, by courtesy of Hooke, Newton had a sound basis on which to build the world-system."[32] Newton combined his new demonstration of the area law with the existing elements of his orbital dynamics manifested in his pre-1669 solution of uniform circular motion. Hooke was the *catalyst*, but Newton was the *creator*.

KEPLER'S LAW OF EQUAL AREAS AND THE POLYGONAL APPROXIMATION

Kepler published his equal area law in 1609, but as a student in the 1660s Newton appears not to have had direct contact with Kepler's works or with other primary astronomical sources, such as those of Ptolemy or Copernicus. Newton derived his early knowledge of astronomical theory from secondary sources in the form of textbooks, many of them inferior works. He employed a variety of equant mechanisms in his early work to predict the speed of the planets in their elliptical paths. It is not clear when Newton first became aware of Kepler's law of equal areas, even as an approximation. One scholar argues that it is possible but unlikely that he knew of the area law as a student.[33] Another, however, argues that there are reasonable grounds for believing that the area law was both known and appreci-

AN OVERVIEW OF NEWTON'S DYNAMICS   25

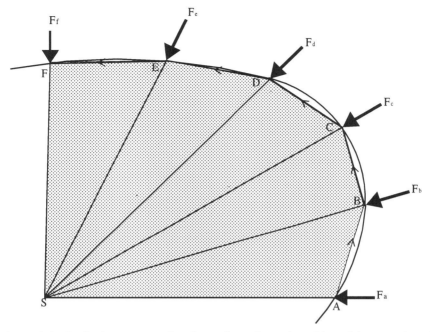

Figure 2.2 If a body moves under the action of a series of impulsive centripetal forces, then it sweeps out equal areas in equal times. The area under the shaded polygon will reduce to the area under the continuous curve as the time between impulsive forces is diminished.

ated by most of the serious astronomers of the period (1650 to 1670).[34] In any event, Newton did not establish the area law as an exact mathematical theorem for a central solar/planetary force until after 1679.

Figure 2.2 is based upon the drawing that appears in the 1684 treatise *On Motion*. Newton used the polygon *ABCDEFS* to approximate the continuous motion of the planet in its path. In this polygonal approximation, the disjointed polygonal path "collides" with the smooth planetary path at a discrete number of points. The motion between any two points on the path, such as *A* to *B*, proceeds in the absence of any force function. A series of impulsive forces F_b, F_c, F_d, . . . , act on the planet only at the discrete points *A*, *B*, *C*, *D*, . . . , where the two paths collide. In general, the forces differ in magnitude, but they are always directed toward the same fixed point *S*. Newton demonstrated that, for a given time between collisions, all the triangular areas *SAB*, *SBC*, *SCD*, . . . , are equal. Thus, equal areas are described in equal times. The law can be demonstrated to be independent of the functional dependence of the impulsive force on distance. The only restriction is that the force always be directed toward the fixed center of

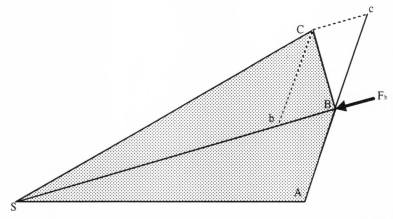

Figure 2.3 The resultant displacement BC is the diagonal of the parallelogram $BcCb$ formed by the virtual displacement Bc and the deviation cC.

force S. The discontinuous motion along the sides of the polygon was ultimately reduced to the continuous motion along the smooth orbital path by letting the size of the triangles, such as SAB, become infinitely small. This limiting process constitutes a hallmark of Newton's dynamics.

As a preliminary review of the tract *On Motion*—which Newton sent to Halley in 1684—consider the following excerpt in which Newton describes the law of equal areas. The details of the proof have been omitted (they are given in chapter 4); for the moment simply follow the flow of the argument.

> *All orbiting bodies describe, by radii having been constructed to their center, areas proportional to the times.*

> *Let the time be divided into equal parts, and in the first part of the time let a body by its innate force describe the straight line* AB. *The same body would then, if nothing impeded it, proceed directly to* c *in the second part of the time . . .*

See figure 2.3 where Bc is the displacement that would have taken place in the given time if the impulsive force F had not acted on the body at point B.

> *Now when the body comes to* B, *let the centripetal force act with one great impulse, and let it make the body deflect from straight line* Bc *and proceed along straight line* BC.

Again, see figure 2.3 where BC is the displacement that did take place in the given time when the impulsive force F did act on the body at point B. The displacement cC is the deviation produced by the change in motion of the body generated by the impulsive force acting at point B. The resulting displacement BC can be seen as the diagonal of the parallelogram formed, with sides given by the virtual displacement *Bc* and the impulsive displacement *cC*.

> *Join* S *and* C *and because of* . . . *the triangle* SCB *will be equal to* . . . *the triangle* SAB.

> *By a similar argument, if the centripetal force acts successively at* C, D, E, *etc., making the body in separate moments of time describe the separate straight lines* CD, DE, EF, *etc., the triangle* SCD *will be equal to the triangle* SCB . . .

Thus, given the demonstration, all the triangular areas in figure 2.2 above are equal (i.e., area *SAB* = *SBC* = *SCD* = *SDE* = *SEF*, etc.).

> *In equal times, therefore, equal areas are described. Now let these triangles be infinite in number and infinitely small, so that each individual triangle corresponds to the individual moments of time, the centripetal force acting without diminishing, and the proposition will be established.*

This result, with the full supporting demonstrations, appears as Theorem 1 in the tract *On Motion* and as Proposition 1 in the *Principia*. Note that Newton did not use the polygonal approximation to obtain the specific dependence of the force upon distance, such as the inverse square law of gravitation. As will be demonstrated in the discussion of impulsive collisions in chapter 3, the specific form of the force is intertwined with both the time during collisions and the time between collisions. Given that the force acts always toward a fixed point, however, it is possible to use the polygonal approximation to predict interesting properties of the force, such as the law that equal areas are swept out in equal times. In his early analysis of uniform circular motion, Newton used the polygonal approximation and the central nature of the force to demonstrate other properties of such motion. Yet, when Newton attempted to find the specific nature of the force (i.e., to solve direct problems) he did not begin with a discrete polygonal path, but rather he developed another approach, one that began with a smooth continuous path. Just as the polygonal approximation is based upon Descartes's dynamics of collisions, so this alternate parabolic approximation is based upon Galileo's linear dynamics.

NEWTON'S DEBT TO GALILEO

Newton was as excessive in offering credit to Galileo as he was in withholding credit from Descartes. The scholium that follows the laws of motion in the *Principia* opens with the following lines:

> *By means of the first two laws and their first two corollaries Galileo discovered that the descent of heavy bodies is in the doubled ratio of the time, and that the motion of projectiles takes place in a parabola, in agreement with experience, except insofar as those motions are slowed somewhat by the resistance of the air.*

While Galileo's direct contribution to "the first two laws" may be questioned, it is clear that his analysis of uniformly accelerated rectilinear motion played an important role in Newton's dynamics. Galileo demonstrated that the motion of a body under a constant acceleration has the following attributes: (1) it is proportional to the square of the time, (2) it can be combined with a uniform motion in another direction, and (3) the combination of such motions gives rise to a parabolic path. All of these elements are found in Newton's earliest writings on dynamics; they appear in the 1684 tract set to Edmund Halley; and they are enshrined as Lemma 10, Corollary 1, and Proposition 6 in the *Principia*.

The proportionality of a body's motion to the square of the time appeared as a hypothesis in the 1684 tract without a demonstration: "The space which a body, with some centripetal force impelling it, describes at the very beginning of its motion, is in the doubled ratio of the time." In the first redrafting of this tract, the hypothesis was promoted to the status of a lemma and a short demonstration was added to the statement given above. It appears in the 1687 *Principia* as Lemma 10 and plays a fundamental role in the demonstrations that follow.

The parallelogram rule also appeared in the 1684 tract as a hypothesis: "A body, in a given time, is carried to the place where it is carried by separated forces in successively equal times." In the first redrafting of this tract, this hypothesis was also promoted to the status of a lemma and a demonstration was added. It appears in the 1687 *Principia* as Corollary 1, which follows the statement of the three laws of motion in a slightly revised form: "A body, with forces having been conjoined, describes the diagonal of a parallelogram in the same time as it describes the sides, with [forces] having been separated."

The parabolic motion, which arises by the combination of an initial projection combined with a constant acceleration, appears implicitly in the 1684 tract in the form of an approximation. The parabolic approximation assumes that at "the very beginning of its motion" the force at any point on a given curve is approximately constant. Thus, the combination of the initial tangential velocity and the acceleration due to that constant force gives rise to a vanishingly small parabolic arc that will approximate the element of the general curve at the given point. No explicit properties of the parabola are employed, but the magnitude of the force is taken as proportional to the radial displacement and inversely proportional to the square of the time. This parabolic approximation appears implicitly in Newton's earliest work and continues unchanged throughout his most mature analysis. It appears in Proposition 6 of the 1687 *Principia* as the basic assumption in the general paradigm for solving direct problems.

Newton's tribute to Galileo implies, however, that the debt extends beyond these three specific results; it implies that credit also is due to Galileo for the first two laws of motion, which Newton gives as follows:

> Law 1. *Every body perseveres in its state of resting or of moving uniformly straight ahead except insofar as it is compelled by impressed forces to change that state.*
>
> Law 2. *A change in motion is proportional to the motive force impressed, and takes place along the straight line on which that force is impressed.*

While there is clear evidence of the influence of Galileo's kinematics of rectilinear motion on Newton's early research, there is no indication that Galileo contributed to the principle of inertia as found in the first law or to the view of the action of impressed force found in the second law.[35] For Galileo, horizontal motion was an approximation to the circular arc of the earth and, as such, any law of inertia related to natural circular motion rather than ideal rectilinear motion. Moreover, Galileo did not demonstrate a clear understanding of "change of motion" as a measure of "impressed force," as set forth in the second law. In any event, it was from the work of Descartes that Newton appears to have obtained the concept of rectilinear inertia.[36]

THE PARABOLIC APPROXIMATION AND THE LINEAR DYNAMICS RATIO

In an early analysis of uniform circular motion, Newton first approximated the circle by using a polygon; then he calculated details of the motion by assuming an intermittent force that acted only at the points at which the polygonal path touched the circle; and then he increased the number of sides of the polygon until the polygonal path approximated the circular path and the intermittent force approximated the continuous force. This early polygonal approximation was eventually employed in the demonstration of the area law, which was discussed above. Figure 2.4 represents such an approximation. The resultant motion along bc is the combination of the tangential displacement by that would have taken place if no collision had occurred at b and the radial displacement bx that would have been produced by the impulse at b if the body had been at rest.

In another early analysis, Newton employed the parabolic approximation rather than the polygonal approximation to analyze the same uniform circular motion. He employed elements of Galileo's kinematics and did not use collisions. In figure 2.5, the circular path is no longer approximated by a polygon, and the force acts continuously rather than intermittently. Critical to the new analysis is Newton's assumption that within the limits of very short time intervals (i.e., as the point x approaches the point b) the force can be assumed to be approximately constant, both in

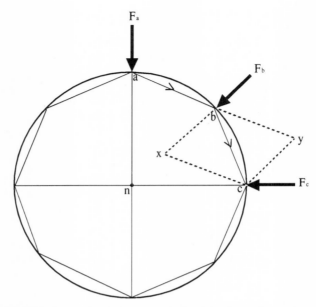

Figure 2.4 The circular path *abc* is approximated by the polygonal path *abc*. After the impulsive force F_b acts at point *b*, the resultant motion *bc* is given by the diagonal of the parallelogram formed by the virtual displacement *by* and the impressed displacement *bx*.

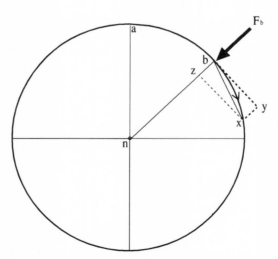

Figure 2.5 The line *by* is constructed tangent to the circle at the point *b*, the line *xy* is constructed parallel to the radial displacement *bz*, and the line *bx* is the diagonal of the parallelogram *byxz*.

magnitude and in direction. Under the action of a constant force alone (i.e., no initial velocity) the displacement would be bz, or its equivalent xy. Under the action of the tangential velocity alone (i.e., no force acting) the displacement would be by. The observed motion along the arc bx is a combination of these two displacements.

Galileo has demonstrated that an initial velocity combined with a constant acceleration gives rise to a parabolic curve, as in ideal terrestrial projectile motion. Here the initial projectile velocity is the tangential velocity along the line by and the constant acceleration is supplied by the approximately constant force and acts along the radial line bn. Whiteside, the editor of Newton's mathematical papers, makes the following observation:

> It will be evident that Newton presupposes that the central force acting upon a body may, over a vanishingly small length of its orbital arc, be assumed not to vary significantly in magnitude or direction, and hence that the infinitesimal arc is approximated to sufficient accuracy by a parabola whose diameter passes through the force-centre, with its deviation from the inertial tangent-line accordingly proportional to the square of the time.[37]

Newton does not speak to this approximation explicitly in the first edition of the *Principia*. In the revised editions, however, he adds the following corollary to Proposition 1 which makes explicit what is only implicit in the first edition. The approximate constant force F is compared to the force of gravity and the deviation xy is called the sagitta (literally the arrow in the bow of the arc).

> Corollary 5. And therefore these forces [the forces F above] *are to the force of gravity as these sagittas* [the deviations xy in fig. 2.5] *are to the sagittas, perpendicular to the horizon, of the parabolic arcs that projectiles describe in the same time.*

Thus, Newton approximates the element of the circular curve in the vicinity of the initial point b by an element of a parabolic curve. The displacement of any future point x on the parabola can be found by using the parallelogram rule to combine the displacement by due to the initial tangential velocity with the deviation xy due to the constant force. The important property of the parabolic approximation, however, is that the magnitude of the force F is directly proportional to the displacement xy and inversely proportional to the square of the time.

As a preliminary review of the *Principia*, consider the following excerpt in which Newton develops his general theorem for a general force function based upon the parabolic approximation. Figure 2.6 is based on the diagram that accompanies Newton's statement of the general theorem. The center of force is located at point S (the sun) and the body is located at point P (the planet). If there were no force acting on the body, then it would travel along the tangent line PR to the point R. Because there is a

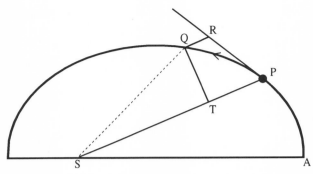

Figure 2.6 The line *PR* is constructed tangent to the curve at point *P*, the line *QR* is constructed parallel to the line *SP*, and the line *QT* is constructed perpendicular to the line *SP*.

force acting on the body, however, it travels along the curve *APQ* to the general point *Q* located beyond the given point *P*. The line *QR* is constructed parallel to the line of force *SP* and thus defines both the point *R* and the displacement *QR*, which is employed in finding the force.

The line *QT* is constructed perpendicular to the line of force *SP*, and the equal area swept out in an equal time is bounded by the lines *SQ*, *SP*, and the arc *QP*. As the point *Q* is brought very close to the point *P*, then the arc *QP* approaches the chord *QP*, and the area swept out is that of the triangle *SQP*, or one-half of the area given by the product of the altitude *QT* and the base *SP*. Thus, the time *t* is proportional to the area *QT* x *SP*.

In the situation described above, as the point *Q* approaches point *P*, the force is assumed to be constant, and the elemental arc of the given curve is approximated by the elemental arc of the parabola, which would be generated by the constant force and the initial tangential velocity at the point *P* (i.e., the parabolic approximation discussed above). The force F, therefore, is proportional to QR/t^2 which in turn is proportional to $QR/(QT \times SP)^2$, or as Newton chose to state it, "The centripetal force is reciprocally as the solid $SP^2 \times QT^2 / QR$."[38] The following is the statement that precedes Newton's demonstration of the theorem as it appears in the 1687 edition of the *Principia*:

[1] *If a body* P *revolving around the center describes any curved line* APQ,
[2] *and if the straight line* [PR] *touches that curve on any point* P,
[3] *and if to this tangent from any other point* Q *of the curve,* QR *is drawn parallel to the distance* SP,
[4] *and if* QT *is dropped perpendicular to the distance* SP;
[5] [then] *I assert that the centripetal force is reciprocally as the solid* SP² x QT²/QR,

[6] *provided that the quantity of that solid that ultimately occurs when the points* P *and* Q *coalesce is always taken.*

Following this statement, Newton provided details of the demonstration wherein the ratio of the displacement and the square of the area $QR/QT^2 \times SP^2$ (i.e., what I have called the linear dynamics ratio) is a measure of the force at the point P in the limit as the point Q approaches the point P. Newton then used this ratio to compute the force for examples of various orbits and force centers. More specifically, it is the ratio of QR/QT^2 that must be expressed as a function of SP. Thus, the ratio QR/QT^2 is the "determinate" of the motion. Newton's exemplar solutions for direct problems can be distilled into the following general pattern of analysis.

Step 1. The Diagram. A drawing is provided that identifies the specific orbit corresponding to the general orbit QPA in the general theorem. The immediate position P of the body is located, and the line of force SP is constructed that connects the body P with the force center S. Then the future position Q of the body is located, and the two lines QR and QT are constructed. Thus, the three elements QR, QT, and SP of the linear dynamics ratio $QR/QT^2 \times SP^2$ are identified in the diagram.

Step 2. The Analysis. Given the full diagram, Newton begins the search for the geometric relationships that will reduce the linear dynamics ratio to a useful form by expressing the discriminate QR/QT^2 as a function of the distance SP alone. It is in this search that Newton displays his command of geometry, conic sections, and mathematical insight; and it is here that the reader may well lose sight of the general structure of the dynamics in the flurry of mathematical details.

Step 3. The Limit. The general theorem holds only in the limit as the future point Q approaches the immediate point P. Thus, Newton need not search for exact geometric relationships, but only for one that will reduce to the desired functional form in that limit. Such relationships will eventually be sought out by others employing the methods of the calculus, but here Newton employs his geometric/limiting technique that serves in its stead.[39]

ELLIPTICAL ORBITS AND THE PARABOLIC APPROXIMATION

After the preparatory propositions of the *Principia*, Newton is ready to produce the solution to the direct Kepler problem. In the first edition, the statement of the direct Kepler problem is given in Proposition 11 as follows:

Let a body be revolved on an ellipse: there is required the law of centripetal force being directed to a focus of the ellipse.

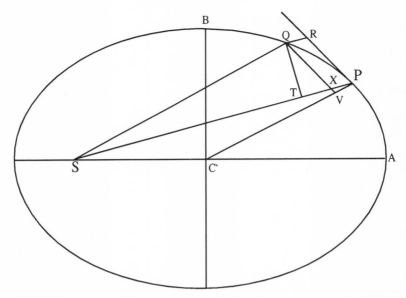

Figure 2.7 The center of force is at point S, the planet is at point P, and point Q is the future position of the planet. As point Q approaches point P the line CV approaches the line CP and the line QX approaches the line QV.

The three steps into which the general pattern of analysis were resolved are now applied to Proposition 11.

Step 1. The Diagram. Figure 2.7 is a diagram that contains some of the elements of the original drawing that pertain to the direct Kepler problem. The curve of the general theorem becomes the ellipse *APQ*, and the immediate position *P* of the body is located on that ellipse. The force center *S* is located at a focus of the ellipse, and the line of force *SP* is constructed. The future position *Q* of the body is also located on the ellipse, and the two lines *QR* and *QT* are drawn, where *QR* is parallel to the line of force *SP* and *QT* is perpendicular to it. The line *QXV* is constructed parallel to the tangent line *PR*.

Step 2. The Analysis. The solution of the problem requires that the discriminate ratio QR/QT^2 be reduced to an appropriate form by expressing it as a function of *SP* and the constants of the ellipse. The direct Kepler problem provides a greater challenge than any of the three preliminary examples that precede it (e.g., the circle, the spiral, and the ellipse with the center of force at the center of the ellipse). In this solution of the direct Kepler problem, Newton employs three specific properties of an ellipse,[40] three geometric properties derived from the construction,[41] and the definition of the *latus rectum*, *L*, a constant of the ellipse.[42] The manipulation of these specific properties is set aside for the moment in order better to understand the general procedure. The penultimate result can then be expressed as follows:

$$QR/QT^2 = (1/L)(CP/CV)(QX/QV)$$

where the line segments *CP, CV, QX,* and *QV* are defined in the diagram in figure 2.7 above.

Step 3. The Limit. The final result, however, is obtained in the limit as required by the general theorem. Thus, "as the point *Q* approaches the point *P*" one can see from the diagram that the line segment *CV* approaches the line *CP* and the line segment *QX* approaches the line *QV.* Therefore, their ratios approach unity and the discriminate ratio QR/QT^2 is proportional to the reciprocal of the constant *latus rectum,* *L*, and hence the discriminate ratio is itself a constant. The force is thus given as follows:

$$F \propto (QR/QT^2)(1/SP^2) \propto (1/L)(1/SP^2) \propto 1/SP^2$$

The force *F* required to maintain an object in an elliptical orbit about a center of force located at the focus of the ellipse depends upon the inverse square of the distance *SP* between the object and the center of force. Kepler had demonstrated that the planet Mars exhibited elliptical motion about the sun located at a focus. Thus, the gravitational force between the sun and a planet depends upon the inverse square of the distance, and Newton set the stage for the new astronomy of the eighteenth and nineteenth centuries.

THE CIRCULAR APPROXIMATION AND THE CIRCULAR DYNAMICS RATIO

There is a sense in which Newton is indebted to Descartes for the analysis underlying the polygonal approximation and there is also a sense in which Newton is indebted to Galileo for some of the analysis underlying the parabolic approximation. The circular approximation, however, is the creation of Newton alone. The parabolic approximation is described above as a process in which the element of the general curved arc of the path in the vicinity of a point is replaced by an element of a parabola generated by the initial tangential velocity and the constant force. The circular approximation, by contrast, is described as the process in which the element of the general curved arc of the path in the vicinity of a point is replaced by an element of the curved arc of the circle that best approximates the general curve at the point (i.e., the circle of curvature or the osculating circle). Newton developed the concept of mathematical curvature and its application to various types of curves in his early work at Cambridge before 1665. Details of how the circular approximation could be applied to

36   THE BACKGROUND

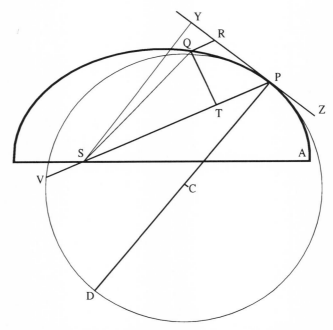

Figure 2.8 The circle *DPV* represents the circle of curvature to the general curve *APQ* at the point *P*. The line *PV* is a chord of the circle and the line *YS* is a perpendicular to the tangent *ZPY*.

direct problems, however, did not appear in Newton's papers until he was preparing revisions for the 1687 edition of the *Principia*, although there are indications that he had considered the general procedure as early as 1664 (see the final section of chapter 3).[43]

Figure 2.8 illustrates the revisions that must be made to the basic figure employed to obtain the linear dynamics ratio in order to obtain the circular dynamics ratio (compare fig. 2.8 to fig. 2.6). The circle *DPV* represents the circle of curvature at the point *P* to the general curve *APQ*; the line *PSV* is a chord of the circle of curvature drawn through the center of force *S*; and the line *YS* is the perpendicular to the tangent to the curve at point *P* also drawn through the center of force *S*. The arc of the general curve in the vicinity of the point *P* is approximated by the arc of the circle of curvature. The known force F_C required to maintain uniform circular motion along the circular arc is the square of the tangential speed v^2 divided by the radius of the circle *PC* and is directed toward the center of the circle of curvature *C*. The unknown force F_S required to maintain the general motion along the arc of the curve is directed toward the general center of force *S*. The two forces are related, and Newton demonstrated that the unknown force is inversely proportional to the product of the square of the perpendicular SY and the chord PV.

$$F_s \propto 1 / (SY^2 \times PV)$$

In this equation, $1/(SY^2 \times PV)$ is what I call the "circular dynamics ratio." Newton did not use this measure of the force in the 1687 edition of the *Principia*, but in the revised editions that followed, in 1713 and 1726, it was employed to develop "alternate solutions" to those obtained by using the linear dynamics ratio. When applied to the Kepler problem of elliptical motion about a focal center of force, the ratio proves to be inversely proportional to the square of the distance SP. Thus, the solution to the direct Kepler problem that was determined by the parabolic approximation is confirmed by the solution that was determined by the circular approximation.

CONCLUSION

Newton produced three distinct but related methods of attacking problems of motion: the polygonal approximation, the parabolic approximation, and the circular approximation. The polygonal approximation is employed in all the editions of the *Principia* to demonstrate Kepler's law of equal areas in equal times. The parabolic approximation is employed in all the editions of the *Principia* to demonstrate that the force is proportional to the linear dynamics ratio (i.e., the ratio $QR/(SP^2 \times QT^2)$). The circular approximation is not in the first edition of the *Principia*, but in all the other editions it provides an alternate solution for direct problems in which the force is proportional to the circular dynamics ratio (i.e., the ratio $1/(SY^2 \times PV)$). Newton's vocabulary, however, may pose problems for the modern reader. Today, a physicist defines the word "force" as "the time rate of change of the linear momentum," and the "linear momentum" is defined as the product of the "inertial mass" and the "vector velocity." When one who is trained to use the word "force" in such a manner encounters Newton's use of the word "force," confusion often ensues. Newton's carefully prescribed manner of using the word "force" emerges from the details of his analysis.[44] What he means by the word, however, is often lost in the glare of modern terminology. Because his choice of terms does not conform to modern usage, Newton has been accused of vacillation or confusion. Richard Westfall, a leading modern scholar, writes:

> Thus *On Motion* attempted to use the parallelogram of motion to compound either of two differing conceptions of force, impulses which produce discrete increments of motion ($f = \Delta mv$) or continuous forces which produce uniform accelerations ($f = ma$), with a third conception of force as an internal propulsion which maintains a uniform velocity ($f = mv$).... The derivation could not have survived critical examination.[45]

The statement above—that Newton attempted to use the parallelogram rule to combine differing conceptions of force—is at best misleading and at worst false. Newton does not use the parallelogram rule to combine forces as one would do today in vector addition. He uses it to combine displacements that measure the forces. The displacements may arise from an impulsive action ($f = \Delta mv$), as in the polygonal approximation; or from a continuous force that produces uniform acceleration ($f = ma$), as in the parabolic approximation; or from an internal propulsion that maintains a uniform velocity ($f = mv$). One may question what, in Newton's view, maintains the motion (the nature of inertia), or the source of the force (the nature of centrifugal force), or even what he calls a force (the nature of impulse), but there is no confusion in Newton's procedure. The derivation can and does "survive critical examination."

Beyond the derivation, it has been suggested that the conceptual development of the second law survives critical examination. Bernard Cohen argues that Newton believed in the continuous sequence of time that nevertheless came in constant "infinitesimal" units.

> For us, in other words, Newton's "impulse" [I] is the limit of ($F \times \Delta t$) for small values of Δt, whereas for Newton, as $\Delta t \to 0$, $I \to F$. That F and I have different physical dimensions was not a problem for Newton, nor did he ever declare explicitly that these two types of "force" differed by an assumed (or "built-in" factor of time), Δt or dt.[46]

Because Newton never deals with specific numerical examples, the constant infinitesimal units of time are of no consequence. Cohen therefore sees no conceptual difference for Newton between the various measures of force.

I am struck, in fact, by how little Newton's basic kinematic and dynamic analysis changes from its beginning in the *Waste Book* and its fruition in the *Principia*. The terminology undergoes a sense of evolutionary development over the years as he continually attempts to set the analysis forth more clearly and explicitly. Moreover, the conceptual structure that surrounds his dynamics undergoes considerable evolution. Nevertheless, his earliest writings on dynamics employ a consistent set of elements of orbital analysis that survive into his more mature writings. As an example of the consistency of analytic method in the midst of conceptual change, consider the following statement by Domenico Bertoloni Meli concerning the role of centrifugal force in Newton's calculations:

> In his maturity Newton interpreted centrifugal force in orbital motion as a reaction to centripetal force: as such, they were considered to be equal and opposite—whilst in the past they appeared to be only opposite but not necessarily equal—*and then centrifugal force was ignored in the calculations*. [emphasis added][47]

Julian Barbour, writing of Newton's evolution of interpretation of the cause of uniform rectilinear motion, comes to a similar conclusion concerning the consistency of the analytic method.

> The young Newton appears to have conceived this [uniform rectilinear] motion very much in the manner of medieval impetus theory, whereas the mature Newton moved more to the Cartesian standpoint. . . . However, this [change in metaphysics] did not alter any of the mathematical consequences of Newton's theory. *The mathematics—and hence the objective content—of an existing theory is indifferent to the metaphysics through which it is interpreted.* [emphasis added][48]

In each of the three approximations, Newton began with the analysis either of collisions, or of combined uniform rectilinear motions, or of uniform circular motion, and he developed from each a measure of force for more complex motions. Each of these approximations has its roots in the very early work that Newton did before 1669 and each appears in the more mature work that he did after 1679. The story must begin with Newton's first thoughts on dynamics in 1665, if we are to understand the full thrust of his later thoughts, and to that end we turn to his early education.

THREE

Newton's Early Dynamics
On Uniform Circular Motion

Isaac Newton was born on 25 December 1642, in the manor house of Woolsthorpe, a very small village seven miles south of the town of Grantham. Fifty miles south of Grantham lies the university town of Cambridge, the center of Newton's university training and scholarly activities. Seventy miles south of Cambridge lies London, the location of Newton's later professional life and of his death on 27 March 1727. Newton exerted an influence, however, that extended well beyond the miles between Grantham and London and lasted much longer than the period of eighty-four years between his birth and death; his published work exerted its authority throughout the entire academic world and its influence is still felt today.

Shortly before Newton's birth, his father died, and three years later his mother married the rector of the church in a neighboring village. Newton did not live with his stepfather and mother but remained at Woolsthorpe in the care of his grandmother for nine years until his stepfather's death in 1653. Newton's mother returned to Woolsthorpe with her three children from her second marriage and, among other items, her husband's library of some two or three hundred books on theology, which may have laid the foundation for Newton's interest in theology. In addition to the books on theology, the library contained a large bound notebook of blank paper, which Newton retained and called his *Waste Book*. His stepfather had intended the bound notebook to serve as a repository for his theological studies but had filled only a few pages. Newton, however, filled many pages with his thoughts on a great variety of subjects, including his first work on dynamics.

Newton left his home in Woolsthorpe to enter the grammar school in Grantham at age twelve, less than two years after his mother returned home. His early mathematical training apparently did not extend beyond the elementary rules of manipulation that were standard for the time. No evidence exists to suggest that he displayed in grammar school any of the extraordinary talent for mathematics that manifested itself shortly after he arrived at Cambridge University.

Newton, at age nineteen, entered Trinity College of Cambridge University in the summer of 1661. His first two undergraduate years were devoted to the standard scholastic curriculum of the time. His grammar school training had provided him with a firm grasp of Latin with which to begin the traditional university program of scholastic philosophy. Newton's notes on his early readings indicate that he began to work through a Latin summary of Aristotle's natural philosophy, probably under the direction of a tutor. He made notes on the sections that summarized or commented on Aristotle's *Physics, On the Heavens, On Generation and Corruption,* and *Meteorology.* He also read portions of a compendium of the main theses of Aristotelian metaphysics. This early study of the Aristotelian body of works gave him a system that organized the diversity of nature into a coherent pattern and stressed rigorous thought. Newton's *Principia* played a major role in the rebellion against scholasticism, but it was the scholastic system itself that initially provided him with a firm foundation from which to rebel.[1]

Although Newton began his college training in the classical tradition, it was not long before he began to look outside the traditional curriculum, which was itself in a state of decline. Academic Aristotelianism was maintained in part by the legal mandate of a curriculum enacted by law and in part by academics with a vested interest in continuing the system that had produced them.[2] The traditional curriculum was no longer a stimulant to intellectual vigor and by 1661 European philosophy had developed a new set of challenges. Thus, it comes as no surprise that a serious and perceptive student such as Newton never finished the Aristotelian texts he was assigned, although these early studies may have continued to influence his worldview. By 1663, the commentaries in his student notebook on the Aristotelian texts were interrupted by pages of notes devoted to the metaphysics of the contemporary French mathematician and philosopher René Descartes. He digested the work of Descartes in a way that he never had done with the work of Aristotle. He did not confine himself solely to Descartes, however. He read of the new mechanical philosophy in the works of the British physicist and chemist Robert Boyle and the British philosopher Thomas Hobbes. Moreover, he read Galileo's *Two New Sciences,* and he familiarized himself with contemporary astronomical writings. By 1664, however, mathematics dominated his thoughts to the virtual exclusion of everything else. In the field of mathematics, Newton was almost entirely self-taught. He attended a few lectures on the subject, but it was his reading of Descartes's *Geometry* and other contemporary works on mathematics that proved his real classroom. In just over a year he had mastered the entire achievement of seventeenth-century mathematical analysis and had begun to make major contributions to it himself.

Newton concerned himself early on with topics that proved to be valuable in his later study of dynamics. In particular, the topic of conic sections was to prove important in his analysis of celestial motion. He collected various methods of constructing conic sections, investigated the role of coordinate systems in simplifying such constructions, and organized his growing knowledge of analytical geometry. During this period of self-study, Newton was creating, as well as absorbing, mathematical procedures. His development of the basic mathematical techniques of his fluxional calculus, for example, reached fruition in the tract of October 1666. Although he did not apply the method of fluxions directly to his dynamics, the work provided experience in taking limits. Newton relied on this experience when he manipulated the quasi-classical geometry that he did apply to dynamics.

By 1666 Newton concluded his first creative drive into the realm of mathematics after almost a year and a half of total dedication. It is almost impossible to overstate the degree of his success. Newton began to make original contributions to the field of mathematics even in the process of taking notes on his reading as a student. Initially, his study of existing mathematics was intermingled with his creation of new mathematics. Within the first year of his mathematical activity, however, his interest in learning from others waned and he became driven by the stimulation of his own intellect. The products of this intense activity, however, remained unknown to members of the mathematical community. Even his academic mentor, Isaac Barrow, did not know the extent of Newton's progress. Newton's research in analytical geometry and calculus from 1664 to 1666, had it been known, would have placed Newton in the front ranks of contemporary mathematicians, and with the October 1666 tract on fluxions he may well have been considered the leading mathematician of all Europe.

UNIFORM CIRCULAR MOTION:
THE POLYGONAL APPROXIMATION

As his first wave of mathematical creativity came to a close early in 1665, other interests began to claim Newton's attention. Among them was his study of dynamics founded on the works of Galileo and Descartes. His first investigations into this subject were concerned with collisions and appear in his bound notebook, the *Waste Book*. The only date among the dynamical entries in the *Waste Book* was the marginalia, "Jan. 20th 1664" (1665 new style),[3] which appeared in a section devoted to problems of collisions between two perfectly inelastic bodies.[4] In this section, Newton developed and refined the concepts and axioms of motion that Descartes had set out in 1644. In particular, Newton's statement of Descartes's concept of linear inertia originally appeared in Newton's *Waste Book* as follows:

> *Every thing doth naturally persevere in that state in which it is unlesse it bee interrupted by some externall cause, hence* . . . *A body once moved will always keepe the same celerity* [*speed*], *quantity, and determination* [*instantaneous direction*] *of its motion.*[5]

This Cartesian view that uniform rectilinear motion is the natural isolated state is crucial to Newton's dynamics, and an enlarged version of the statement appeared some twenty years later as the first law of motion in the *Principia*.

Newton extended the dynamics of intermittent colliding bodies to his analysis of bodies subject to a continuous force. He determined the nature of the force directed toward the center of a circle that will move the body uniformly around the circle (i.e., the solution to the direct problem of uniform circular motion). Newton envisioned a ball in uniform circular motion about a given center, and he sought information concerning the nature of the force impressed upon the ball. In a later work he demonstrated that force was directly proportional to the square of the speed and inversely proportional to the radius of the circle. In the section on uniform circular motion that first appears in the *Waste Book*, however, he did not explicitly obtain that result. The relationship he did obtain compared the magnitude of the change in motion produced by a sum of impulsive forces to the magnitude of the orbital motion. It is not the type of problem that would concern a contemporary physicist. Newton's analysis, however, and not the specific problem, is of interest.

In this analysis the circular path is approximated by a polygonal path of n sides (shown in fig. 3.1 as an octagon), where the number of sides of the polygon will ultimately increase without limit. The ball is envisioned to collide elastically (i.e., the speed or magnitude of the velocity does not change) with the circle or with the circumscribed polygon. At each collision, an impulsive force, directed toward the center of the circle, acts on the ball; between collisions no force acts on the ball and it moves with uniform rectilinear motion. The impulsive force acts for an extremely short time and produces a change in what Newton called the "quantity of motion in a body." He calculated the sum of the magnitudes of these changes in the quantity of motion, which he called "the force of all the reflections," and compared that sum to the quantity of motion of the body's constant orbital motion, which he called "the force of the body's motion." As the number of sides of the polygon was increased, the polygonal path approached a circular path, and the ratio of "the force of all the reflections" and "the force of the body's motion" was found to approach the value of 2π.

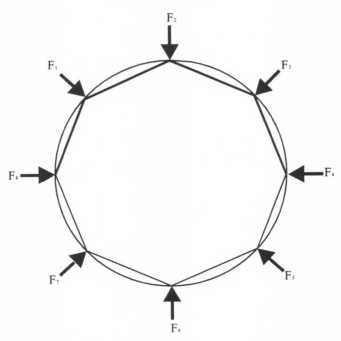

Figure 3.1 A circular path is approximated by an octagon. An impulsive force F acts on the body at each intersection of the circle and octagon.

To the modern eye, Newton's demonstration may seem insufficient because he did not calculate the mathematical structure of the force, directed toward the center of the circle, that is required to maintain uniform circular motion. Moreover, the modern reader will find that Newton's terminology is at variance with current usage. In modern terminology, the change in motion that Newton calculated at each collision is the magnitude of the change in the linear momentum Δmv. In the terms used in a modern textbook, this quantity Δmv is equal to the *product* of the force F and the small time of collision δt, where the product $F \times \delta t$ is defined as the impulse I. The magnitude of the vector linear momentum mv does not change at an elastic collision, only its direction. Hence, there is a finite change in the vector linear momentum Δmv provided by the impulsive force. There is no explicit use of such a modern vector resolution in Newton's text, but it resides implicitly in his analysis. One must not judge the value of the text from the viewpoint of modern expectations. Newton was exploring the phenomenon of uniform circular motion, and the techniques he developed in that exploration served him as a springboard for further analysis. The methods of analysis found in this early text provided a basis to continue into the more mature considerations to follow.

The text of Newton's analysis, which was written in English, is found on the first pages of the *Waste Book* and is dated about 1665.[6] Because of the complicated but polished nature of the text, it has been argued that this passage may not represent Newton's first written thoughts on the subject of circular motion, even though it appears first in the book. He may have worked out the demonstration on loose sheets, revised it, and then transcribed it onto the blank first page.[7] Newton makes an opening formal statement of the result he intends to demonstrate, and then he gives his detailed demonstration or proof of that opening statement. Below is this opening statement, simply intended to set forth the problem and solution. It will be followed by a detailed mathematical demonstration.

UNIFORM CIRCULAR MOTION: THE *WASTE BOOK*

If the ball b *revolves about the center* n, [*then*] *the force by which it endeavors from the center* n *would beget soe much motion in a body as there is in* b *in the time that the body* b *moves the length of* [*an arc equal to*] *the semidiameter* bn.[8] *Or the force from* n *in one revolution is to the force of the body's motion as* :: *periph* : *rad.*

Note that, following Descartes, Newton is concerned with the "force by which it [the ball] endeavors from the center." That Cartesian outward endeavor will be identified below with the "force or pression" of the ball *b* upon the curve "at its reflecting." Some fifteen years later, after Newton had demonstrated the law of equal areas and had challenged the Cartesian system of mechanical vortices, the specific reference will be to the "center-seeking force" rather than to the outward endeavor. As will become clear in the detailed analysis in the next section, the "endeavor from the center" will be measured by the difference between the tangential displacement the ball would have made in the absence of the force and the displacement it does make in the presence of the force. That difference between the two displacements is itself a radial displacement, one which is produced by a radial force. The details of the analysis, however, are independent of Newton's identification of the displacement as either an outward or an inward displacement.

What the young Newton set forth in his opening statement (perhaps in not too clear a fashion) are two specific examples of the effect generated by the force required to produce uniform circular motion. He demonstrates these examples in the text that follows the opening statement.[9] I paraphrase them as follows:

First Example. If the body traverses a circular arc of length S_1 equal to the radius (semidiameter) of the circle *bn*, then the "motion in a body" generated (begat) by the force of "endeavor from the center" is to the "force of the body's motion" as the arc length S_1 (= *bn*) is to the radius *bn* (i.e., as 1 is to 1).

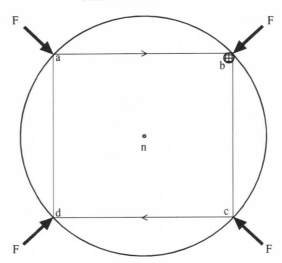

Figure 3.2 A circular path is approximated by a square path. An impulsive force *F* acts on the body at each intersection of the circle and square.

Second Example. If the body traverses a circular arc of length S_2 equal to the circumference (one revolution) of the circle $2\pi bn$, then the "motion in a body" generated by the force of "endeavor from the center" is to the "force of the body's motion" as S_2 (= $2\pi bn$) is to the radius bn (i.e., as 2π is to 1).

In what follows, Newton considers three specific variations: first, motion along a square path inscribed in a given circle; second, motion along a general polygonal path inscribed in the same circle; and third, motion along the circular path itself. In the first example, the ratio of (A) the sum of "the force of reflections" at the four corners of the square to (B) the "force of the body's motion" along one of the sides of the square is found to be equal to the ratio of (C) the sum of the sides of the square to (D) the radius of the circumscribed circle (i.e., A/B = C/D). Then he extends the result to a general polygonal path. Finally, the number of sides of the polygon is increased without limit and the result is obtained for a smooth circular path.

Newton began his demonstration of the effect of the force required to maintain uniform circular motion by considering the motion of a body contained in a square, as shown in figure 3.2. The force is described as "the force of reflection," and the body (i.e., the "Globe *b*") is envisioned to "collide" with the circle or circumscribed square at points *a*, *b*, *c*, and *d* where it is acted upon by an impulsive force *F*. The force on the body is supplied by the collisions of the body with the sides of the circle. Elsewhere in the *Waste Book*, Newton discussed the uniform circular motion of a sphere as it rolls around inside a physical cylinder. The source of the force was the contact between the sphere and the wall of the cylinder.[10] In the demonstration above, however, the circle is not explicitly described as a material body (such as the cylinder). Because the body's free motion is along the tangent line, and because the body deviates from the tangent along a chord of a concave circular arc, then there must be an inward impulsive radial force supplied by some external agency. The source of the inward radial impulse is the elastic collision of the body with the circle. The concern with the force produced by a collision is the hallmark of Cartesian dynamics, and it was perhaps natural that the young Newton began his analysis of dynamics with the interactions described. In what follows, each line of Newton's demonstration will be considered separately.

Square

[1-A] *If* ef = fg = gh = he = 2fa = 2fb = 2gc = 2ed. *And* [*if*] *the globe* b *move from* a *to* b

See figure 3.3 for an inscribed and a circumscribed square in and around the circle given above. The body collides with the circle only in the middle of each side of the circumscribed square, and between collisions the body moves with uniform rectilinear force-free motion (i.e., inertial motion). Line [1-A] simply states that the sides of the circumscribed square (*ef, fg, gh,* and *he*) are bisected by the points *a, b, c,* and *d,* and thus *ef* = 2*fa,* and so on.

[1-B] *then* 2fa : ab :: ab : fa :: *force or pression of* b *upon* fg *at its reflecting* : *force of* b's *motion.*

Or, $2fa/ab = ab/fa$. This result may be obtained from the Pythagorean theorem applied to the right triangle *afb* in figure 3.3: $ab^2 = fa^2 + fb^2 = 2fa^2$, where $fa = fb$. Thus, $2fa^2 = ab^2$ or, dividing by *fa* and *ab*, one has $2fa/ab = ab/fa$, as required in the first proportion.

The second proportion requires that the ratio of the "force or pression" to the "force of motion" equals $2fa/ab$ (and hence $= ab/fa$). The "force of motion" is proportional to the body's initial velocity, and thus in a given time it is proportional to the distance the body would travel with that initial velocity (i.e., the distance *ab*). The "force or pression" is proportional to the change in velocity of the body that the impulsive force would produce, and thus in a given time it is proportional to the distance 2*fa* generated by that change in velocity. Implicit in this statement is Newton's understanding that the impulsive force exerted by the body *b* on the side of the square *fg* at a reflection is equal to the impulsive force exerted on the

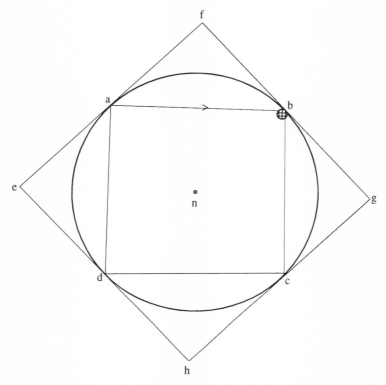

Figure 3.3 A circle with an inscribed and circumscribed square. A body rebounds at points *a, b, c,* and *d* and travels along the inscribed square.

body by the square (i.e., action and reaction). Newton made this relationship explicit when he reworked this example for inclusion in the *Principia* (see the "alternate demonstration" to follow this demonstration).

The ratio given in [1-B] entails an implicit relationship between force and displacement that is fundamental to Newton's dynamics: the force at a reflection is measured by the deviation from the rectilinear motion that the reflection produced. This relationship between force and displacement is employed in all of Newton's subsequent analyses of direct problems, but he uses it here without an explicit development. Whatever role the outward endeavor plays in Newton's thoughts about the nature of the force, here it is only the inward radial displacement, and hence the inward radial force, that is measured.

In another tract, dated 1666, Newton presented an explicit parallelogram rule for combining displacements (see fig. 3.4), and it was this rule

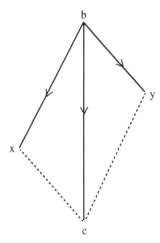

Figure 3.4 The parallelogram rule: the displacement *bx* combined with the displacement *by* is equivalent to the displacement *bc*.

that he implicitly applied in [1-B]. The rule states that if the displacement *bx* from point *b* to point *x* is combined with the displacement *by* from point *b* to point *y*, then the net effect in a given time is the displacement *bc* from point *b* to point *c*, where the point *c* is generated by completing the parallelogram *bycx* (i.e., *bc* is the diagonal of the parallelogram formed with *bx* and *by* as two adjacent sides).[11]

This parallelogram rule can be applied to figure 3.5, in which a point *y* is defined and a number of lines are added to Newton's original figure. The two displacements *by* and *bd* do not actually take place because they are virtual or imagined displacements. The two virtual displacements *by* and *bd* are combined and the resultant displacement *bc* is obtained:

1. The virtual displacement *by*, which would have been produced in a given time by the initial uniform motion the body had before the collision at point *b* and which, if the collision had not taken place, would have carried the body to point *y*.
2. The virtual displacement *bd*, which would have been produced by the change in motion due to the inward radial impulsive force *F* acting at point *b* and which, if there were no initial motion, would have carried the body to point *d*.
3. The actual displacement *bc*, which was the diagonal of the parallelogram *bycd* formed with one side as the line *by* and the other side as the line *bd*, did carry the body to point *c*. The parallelogram is shown in figure 3.5 as the shaded area.

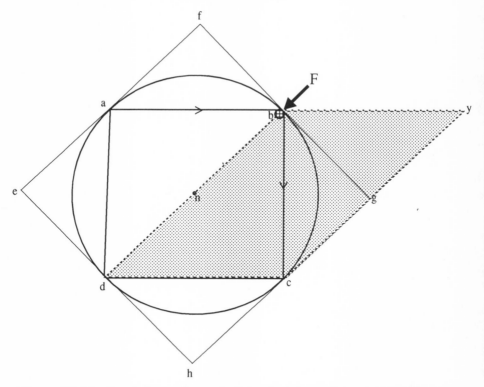

Figure 3.5 The line *bc* is the diagonal of the shaded parallelogram *bycd*. The line *bd* is the deviation due to the impulsive force *F* acting on the body at the point *b*.

The actual displacement *bc* can thus be viewed as the resultant of the virtual displacement *by*, due to the initial motion, and the virtual displacement *yc* (equivalent to *bd*), due to the change in motion caused by the impulsive force. The line *yc* can be seen, for a given time, as the deviation of the body from the rectilinear path *by* that it would have followed if the force had not acted at point *b*. For a given time, the deviation *yc* is directly proportional to the impulsive force and thus is a measure of that force.[12] Moreover, *yc* is equal to 2*af*. (See fig. 3.5, where *yg* = *gc*, and *gc* = *af*, and thus *yc* = 2*gc* = 2*fa*.) Thus, combining the statements, as Newton does without demonstration, the ratio of the "force or pression" to the "force of motion" equals the ratio of the [displacement *yc* to the displacement *by*] which equals 2*fa* / *ab* = *ab*/*fa*.

Newton's use of the word "force" in the context of the collisions just discussed, however, is not consistent with modern usage. In contemporary terms, the impulse *I* is defined as the product of the force *F* and the time interval δt, and this product is demonstrated to be equal to the change in momentum Δmv (magnitude and/or direction). In these modern terms, Newton's "force of reflection" is best rendered as the "impulse" or the change in momentum produced by the impulsive force F acting over the very small time δt of the collision. Expressed in terms of the impulse I_b ($= \Delta mv$) at point b and the initial impulse I_o ($= mv$), the ratio of the "force or pression" to the "force of motion" is equal to the ratio $I_b/I_o = ab/fa$.[13]

[1-C] *Therefore* 4ab = ab + bc + cd + da : fa :: *force of reflection in one round* (*viz.*: *in* b, c, d, *and* a) : *force of* b*'s motion.*

The statement in [1-C] is an extension of [1-B] in which it was demonstrated that for a single collision the ratio of the "force or pression" to the "force of motion" is equal to the ratio of the deviation/displacement which equals $yc/bc = 2af/ab = ab/fa$. In [1-C] Newton summed all of the "force or pression" in one complete trip around the square (i.e., the scalar sum of the four impulses at points a, b, c, and d), and he expressed it in ratio to the "force of motion." The ratio is now the sum of $ab/fa + bc/fa + cd/fa + de/fa$, or, since $ab = bc = cd = de$, the sum can be expressed as $4ab/fa$. Thus, $4ab/fa$ is equal to the ratio of the "force of reflection in one round" to the "force of b's motion" which equals $\Sigma I_b / I_o$, where ΣI_b is the sum of the four impulses exerted in "one round" and I_o is the initial impulse (i.e., $I_o = mv$).

The statement in [1-C] concludes the discussion of the motion on the square. The ratio of the total "force of reflection in one round" to the "force of b's motion" is found to be equal to the ratio of the periphery of the square ($4ab$) to the radius of the circumscribed circle ($fa = bn$). In what follows, this result will be extended first to a polygon and finally to a circle.

Polygon

[2-A] *By the same proceeding, if the globe* b *were reflected by each side of a circumscribed polygon of 6, 8, 12, 100, 1,000 sides, etc.*

Or, $I_b / I_o = ab / na$ (to be demonstrated). Now, the square path is replaced by a general polygonal path. See figure 3.6 for an example using an octagonal polygon. The deviations xb and yc are constructed parallel to the direction of the impulses I_a and I_b (arising from reflections at points a and b), and they are thus parallel to radii na and nb, respectively. Recall that the body rebounds from the circle (or from the circumscribed polygon) at point a and continues with its speed unchanged toward point b. Point x is defined as the position the body would have reached, in the same time as it traveled to point b, if the impulsive force had not acted at point a. Because the time and the speed are the same over the distance xa and ab

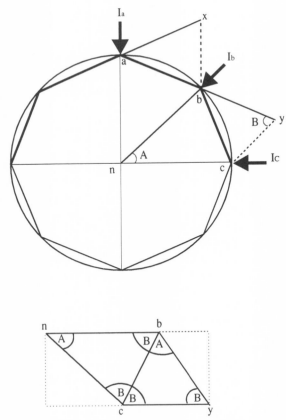

Figure 3.6 The circle is approximated by an octagonal path where the deviations *xb* and *yc* are due to the impulses I_a and I_b at points *a* and *b*. Note that the triangles *bnc* and *ybc* are similar (see bottom figure).

(although the directions differ), then $xa = ab$. Point *y* is related to point *c* in the same way: $yb = bc$.

Note also that triangles *bnc* and *ybc* are similar. See figure 3.6 where *nb* is parallel to *yc*, $nb = nc$, and $bc = by$: thus, the apex angles *A* are equal, as are the base angles *B*, and triangles *bnc* and *ybc* are therefore similar because all the angles are equal. From that similarity, the ratio of the sides of the triangles *bnc* and *ybc* can be written as $yc/bc = bc/nb$. Furthermore, note that figure *anbx* is identical to figure *bncy*, and thus $xb/ab = ab/na$. Hence, since $I_b / I_o = xb/ab$, then $I_b / I_o = ab/na$, which is a generalized relationship for any polygon and therefore contains the relationship in [1-B] for a square (i.e., $I_b / I_o = ab/fa$).

[2-B] [*then*] *the force of all the reflections is to the force of the body's motion as the sume of those sides to the radius of the circle about which they are circumscribed.*

$\Sigma(I_b) / I_o$ (to be calculated). This ratio is the scalar sum of all the impulses $\Sigma(I_b)$ required to bring the body back to its starting point (i.e., "the force of all the reflections") divided by the initial impulse I_o required to set the body into its orbital motion (i.e., "the force of the body's motion").

$\Sigma(I_b) / I_o = \Sigma(ab) / na$ (to be demonstrated). From [2-A], $I_b / I_o = ab/na$, where na is the radius of the circle about which the polygon is circumscribed. Thus, summing over both ratios, the sum is given as $\Sigma(I_b / I_o) = \Sigma(ab/na)$, or, since both I_o and the radius na are independent of the sum, $\Sigma(I_b) / I_o = \Sigma(ab) / na$, where $\Sigma(ab)$ is the scalar sum of all the sides of the polygon or "the sume of those sides," and na is "the radius of the circle about which they are circumscribed."

The statement in [2-B] concludes the discussion of the motion on the general polygon. The ratio of the total "force of all the reflections" in one round to the "force of b's motion" is found to be equal to the ratio of the periphery of the polygon (Σab) to the radius of the circumscribed circle (na). This result is similar to that given for the square in [1-C] above and for the circle in [3-B] below.

Circle

[3-A] *And so if the body were reflected by the sides of an equilateral circumscribed polygon of an infinite number of sides (i.e., the circle itself)*

If the number of sides of the polygon increases without limit, then the perimeter of the circumscribed polygon, $\Sigma(ab)$, approaches the circumference of the circle, $2\pi(na)$.

[3-B] [*then*] *the force of all the reflections is to the force of the body's motion as all those sides (i.e., by the perimeter) to the radius.*

Or, $\Sigma(I_b)/I_o = \Sigma(ab)/na \to$ (circumference) / (radius) (to be demonstrated). From [2-B], the ratio of the scalar sum of the magnitudes of all the impulses in one complete revolution $\Sigma(I_b)$ is to the magnitude of the initial impulse required to set the body into its orbital motion I_o as the scalar sum of all the displacements $\Sigma(ab)$ is to the radius of the circle na. From [3-A], the sum of the sides $\Sigma(ab)$ approaches the circumference of the circle $(2\pi)(na)$ as the number of sides of the polygon increases. Thus, "the force of all the reflections" $\Sigma(I_b)$ is to the "force of the body's motion" I_o as the circumference of the circle is to its radius, or the ratio $\Sigma(I_b)/I_o = $ (circumference)/(radius) $= (2\pi)(na)/(na) = 2\pi/1$. Thus, as set forth in the opening statement, "the force from n in one revolution is to the force of the body's motion as :: periph : rad."

From the vantage point of contemporary physics, the result of this seventeenth-century text is almost trivial. The scalar sum of the changes in the linear momentum $\Sigma(mv)$ is given by the product of the constant magnitude of the circular force $F = mv^2/r$ and the period of revolution $T = 2\pi r/v$ (i.e., $\Sigma(mv)/mv = FT/mv = (mv^2/r)(2\pi r/v)/mv = 2\pi$), as Newton demonstrated above.[14] Nevertheless, the polygonal approximation, which is central to this analysis, is a useful and important method as demonstrated. It was employed by Newton in his later dynamics to demonstrate Kepler's law of equal areas.

AN ALTERNATE DEMONSTRATION

Newton reworked this example as an alternate demonstration in the scholium of Proposition 4 in the 1687 *Principia*, and there he did obtain the nature of the force directly.[15] His motivation for resurrecting this pre-1665 demonstration appears to stem from his desire to establish that he developed his method before Robert Hooke's suggestions in 1674. Proposition 4 demonstrates that the force directed toward the center of a circle required to maintain uniform circular motion is "as the squares of the arcs described in the same time divided by the radii of the circles." The scholium to the proposition offers the following variant demonstration in which the polygonal approximation is employed to obtain the nature of the force directly:

> [A] *The preceding [statements] can also be demonstrated in this way. In any circle let a polygon of any number of sides be described. And if a body, in moving at a given velocity along the sides of the polygon, is reflected from the circle at each angle of the polygon, [then] the force with which it impinges on the circle at each reflection will be as its velocity. And thus the sum of the forces in a given time will be as that velocity and number of reflections conjointly.*

The critical assumption is that the impulsive force F at a collision is proportional to the velocity v. Newton does not attempt to defend the statement in the 1684 revision, nor is it true for a single collision. As demonstrated above, it is necessary to discuss the interactions in terms of the impulse I_b, which is defined as the product of the force F and the time of collision δt. In statement [2-A] in the pre-1665 version above, a generalized relationship $I_b/I_o = ab/na$ is demonstrated for any polygon, where ab is a leg of the given polygon, na is the radius of the circle, and I_o is mv. The impulse I_b is equal to $[(ab/na)m]v$. Thus, it is the impulse I, and not the force F, that is proportional to the velocity v. The sum of the impulses over a given time T is $\Sigma I = NI$ and is proportional to Nv, where N is the number of collisions in that given time.

> [B] *that is, (if the species of the polygon is given) as the [product of the] length described in that given time and that length divided by the radius of the circle: that is, as the square of that length divided by the radius.*

The total length L traveled in a given time T is proportional to the velocity v and hence to the impulse I. The number of collisions N in the length L is equal to L/l, where l is the length of one of the sides of the given polygon. That length l is proportional to the radius of the circle R and thus the number of collisions is proportional to L/R. Thus, the sum of the impulses is proportional to Nv, and since $N \propto L/R$ and $v \propto L$, the sum of the impulse is proportional to L^2/R.

> [C] *Thus, if the polygon, with its sides diminished infinitely, coincides with the circle, then the sum of the forces in a given time will be as the square of the arc in a given time is divided by the radius.*

Now, as the sides of the polygon diminish infinitely, the sum of the impulses over a given time T is $\Sigma I = \Sigma F \delta t = F \Sigma \delta t = FT$, where $\Sigma \delta t = T$ and the force is now continuous. From [B], ΣI is proportional to L^2/R, and since T is given, then it is correct to say that the force F is proportional to L^2/R. Since the length L becomes equal to the arc of the circle as the polygon approximates the circle, then the continuous force F is proportional to "the square of the arc in a given time divided by the radius."

> [D] *This is the [centrifugal] force by which the body urges the circle, and equal to this is the opposite force by which the circle continuously repels the body toward the center.*

The early statement of the problem was framed in terms of the force exerted by the body on the circle during the collisions. Newton now makes explicit the relationship that was left implicit: the forces are equal and opposite. In the first corollary to this proposition Newton notes that the arc is proportional to the velocity and hence the force is proportional to the squares of the velocities divided by the radius of the circle. Thus, the polygonal technique can be extended to obtain the nature of the force. (It is of interest to note that in the first edition of the *Principia* the word "centrifugal" is not explicitly used in the statement, but it is inserted into all the later editions.) Another technique, however, proved to be even more useful in solving direct problems: the parabolic approximation. Its roots also lie in Newton's early work and are carried forward into his mature dynamics.

UNIFORM CIRCULAR MOTION:
THE PARABOLIC APPROXIMATION

The text to be discussed below was written by Newton sometime before 1669, and it explicitly demonstrates that the force necessary to provide uniform circular motion is proportional to the square of the speed and inversely proportional to the radius (i.e., $F = (m)v^2/r$). It is an important text because it demonstrates that Newton used the parabolic approximation technique before 1669. It also gives evidence of the early influence of Descartes's writing upon Newton. More important, however, is the evidence of Newton's dynamic technique of analysis that is independent of—but obscured in part by—the Cartesian terminology used to express it. Newton will continue to employ the parabolic approximation throughout his mature dynamics.

Figure 3.7 is a comparison of the diagram on the left, which accompanied Newton's tract on uniform circular motion, with the diagram on the right, which accompanied the tract from the *Waste Book* discussed above. The obvious difference is that the circular path is no longer approximated by a polygon. In the previous tract, the distance *bx* is related to the impulse produced by the collision at point *a*, and the effects are summed as the number of sides of the polygon are increased. Finally, the limit is considered as the number of sides tends to infinity and the polygon tends to the circle. In the new text, the analysis begins with the consideration of the limit as the point *D* shrinks back to the point *A*. In that limit, the force that produces the displacement *DB* is assumed to be constant in both magnitude and direction. Galileo has demonstrated that such a constant force is directly proportional to the displacement and inversely proportional to the square of the time. The challenge therefore is to express that time in terms of meaningful parameters of the uniform circular motion. In this pre-1669 tract, Newton employs the property that the radius of the circle sweeps out equal *angles* in equal times plus a single proposition from Euclid to demonstrate that the magnitude of the force is proportional to the square of the period of uniform motion and inversely proportional to the diameter of the circle. In his post-1679 work on nonuniform motion, he will also invoke an initial limiting condition in which the force is assumed to be proportional to the deviation from the tangent and inversely proportional to the square of the time (i.e., the parabolic approximation). By this later period, however, Newton has demonstrated that the radius of *any* centripetal force sweeps out equal *areas* in equal times. This result plus an assortment of mathematical propositions and relationships enables him to solve a variety of direct problems, such as determining the force required for elliptical and spiral motions for various centers of force.

The use of the term "parabolic" to describe the approximation employed

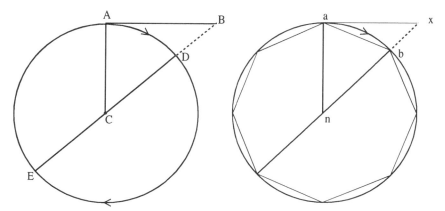

Figure 3.7 A comparison of the diagram employed in the parabolic approximation (left) with the diagram employed in the polygonal approximation (right).

in this 1669 tract stands in contrast to the term "polygonal" used to describe the approximation in the 1665 tract. In the 1665 tract the choice of terms came from the polygonal path. In the 1669 tract the choice of terms comes from Galileo's demonstration that the combination of uniform rectilinear motion and uniformly accelerated rectilinear motion gives rise to a parabolic path. Thus, the element of the circle in the vicinity of the point A can be approximated by an element of a parabola, where the initial projection velocity is the tangential velocity at A and the constant acceleration is given by the approximately constant radial force. No explicit use is made of the properties of a parabola, but the term "parabolic" represents what is implicit in the approximation. Then, from the constant radial acceleration the force is given as directly proportional to the displacement and inversely proportional to the square of the time. This parabolic approximation is carried from this early solution into all of Newton's later work.[16]

It is instructive to consider the outline of Newton's solution in this early tract before looking at the details. From the parabolic approximation the force is assumed to be proportional to the displacement DB and inversely proportional to the square of the time t required to traverse the arc AB. In order to render that relationship in more meaningful terms, Newton's first step is to express the displacement DB and the time t in terms of the diameter DE of the circle and the period T of the uniform motion. In the second step, he calculates the distance X that the body would move in the time T under the influence of the same constant force that produced DB in time t. For a given force, therefore, $X/BD = T^2/t^2$. (See fig. 3.8.) Moreover, because the radius sweeps out equal angles in equal times, the times

58   THE BACKGROUND

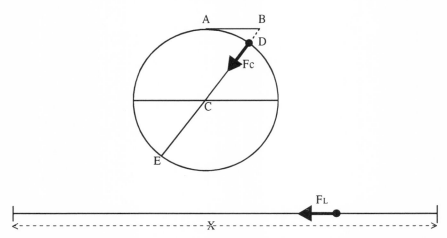

Figure 3.8 Under the action of a constant linear force F_L equal in magnitude to the centripetal force F_C, a body travels the linear distance X in a time T equal to the period of the circular motion.

are proportional to the arcs (i.e., T/t = circumference/arc AD). Calling upon a proposition from Euclid and noting that in the limit the arc DA approaches the tangent BA and the line BE approaches the diameter DE, Newton demonstrates that the distance X is proportional to the diameter of the circle DE. In the third and final step, Newton simply notes that the force is therefore proportional to the diameter and inversely proportional to the square of the period (i.e., $F \propto X/T^2 \propto DE/T^2$) or what is equivalent, $F = kv^2/r$. In what follows, I discuss in detail Newton's derivation of this result.

The paper containing this alternate solution to the problem of uniform circular motion was written some time before 1669. It is not in the *Waste Book* but it survives in another collection of Newton's early writings. Most of the paper is concerned with numerical calculations of the endeavor of the moon from the earth and of the earth from the sun. In the opening section, however, Newton derives the nature of the force in order to make these calculations. The following is an English translation of a portion of the Latin text taken from the manuscript here called *On Circular Motion*.[17] I have imposed the steps, which I summarized in the previous paragraph, upon Newton's solution in the following.

ON CIRCULAR MOTION

Step 1. It is Newton's intent to express the displacement DB and the time t in terms of the diameter DE of the circle and the period T of the uniform motion.

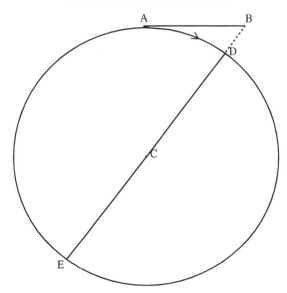

Figure 3.9 A body travels uniformly along the circular path *ADE*. The radial displacement *DB* is the deviation of the circular arc *AD* from the tangential line *AB*.

[1-A] *The endeavor of body* A, *rotating on circle* AD *toward* D, *from the center* [C] *is as great as would carry it away from the circumference to the distance* DB *in the time* AD (*which I set to be most minute*); *inasmuch as it would reach that distance in that time if only it could move freely in tangent* AB *with no impediment to the endeavor.*

See figure 3.9. As noted in chapter 2, two possible readings can be made of this statement. The first identifies the "endeavor from the center" as that which would produce the radial displacement from point *D* to point *B* along the extension of the diameter of the circle. In that reading the "impediment to the endeavor" prohibits the outward radial displacement *DB* and the body travels along the circular arc *AD*. An alternate reading identifies the "endeavor from the center" as that which produces the linear displacement from point *A* to point *B* along the tangent. In that reading the "impediment to the endeavor" prohibits the tangential displacement *AB* and again the body travels along the circular arc *AD*. In either case it is the displacement *DB* that measures the "impediment," be it an outward endeavor or an inward centripetal force. From the full text of this early work, it is clear that Newton intends the first reading (i.e., that expressed by Descartes). In a later version of this solution, produced after 1679, Newton intends the second reading.

[1-B] *Now since this endeavor, provided it were acting in a straight line in the manner of gravity, would impel bodies through spaces that would be as the squares of the times,*

60   THE BACKGROUND

In [1-A], the time to travel the arc AD is "set to be most minute." In that limit of vanishingly small time the force is assumed to be constant in both magnitude and direction "in the manner of gravity." As Galileo has demonstrated, the displacement under a constant force is "as the square of the time."

> [1-C] *In order to find out through how much space in the time of one revolution* ADEA *they would impel [the bodies], I look for a line* [X] *that is to* BD *as is the square of the circumference* ADEA *to* AD^2.

Or, $X/BD = T^2/t^2 = ADEA^2/AD^2$. The line X is the distance that the body would travel under a constant force in a time T equal to the period of circular motion. Since the period T is proportional to the circumference of the circle ADEA (i.e., equal angles in equal times), then the distance X is proportional to the "square of the circumference ADEA." From [1-B], we understand that just as the distance X is proportional to the square of the time T^2 so the distance BD is proportional to the square of the time t (i.e., the arc AD^2). Thus, the relationship, as given by Newton, is $X/BD = T^2/t^2 = ADEA^2/AD^2$.

Step 2. Newton now calculates the distance X that the body would move in the time T under the influence of the same constant force that produced DB in time t.

> [2-A] *To be sure there is* BE : BA :: BA : BD (*by 3* Elem.).

Or, $BE/BA = BA/BD$. This relationship is demonstrated by reference to Proposition 36 of Book 3 of Euclid's *Elements,* a particular Euclidian proposition that Newton employs on a number of occasions, often without a specific reference as given here (i.e., "by 3 *Elem.*"). Figure 3.10 displays the results of Euclid's demonstration: the ratio of the line BE (the diameter plus the deviation) to the line BA (the tangent) equals the ratio of the line BA (the tangent) to the line BD (the deviation). Thus, one has BE : BA :: BA : BD or $BE/BA = BA/BD$.

> [2-B] *Or since the difference between* BE *and* DE, *as also between* BA *and* DA, *is supposed infinitely small, I substitute one for the other in turn and there emerges* DE : DA :: DA : DB.

Or, $DE/DA = DA/DB$. In the limit as the point D approaches A (see fig. 3.11), the extended line BE approaches the diameter DE (i.e., BE → DE), and the tangent BA approaches the arc DA (i.e., BA → DA). From [2-A], $BE/BA = BA/BD$. Substituting DE for BE and DA for BA, that ratio becomes $DE/DA = DA/DB$, as required here in [2-B], or in its equivalent form $DA^2 = (DE)(DB)$, as required next in [2-C].

> [2-C] *Finally by making* DA^2 (*or* DE x DB) : ADEA^2 :: DB : ADEA^2 / DE,

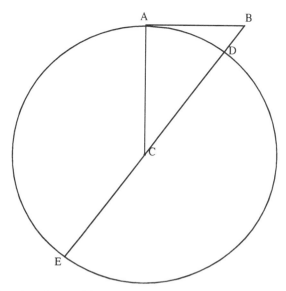

Figure 3.10 Based on Proposition 36 of Book 3 of Euclid's *Elements*: $BE : BA :: BA : BD$ or otherwise $BE / BA = BA / BD$.

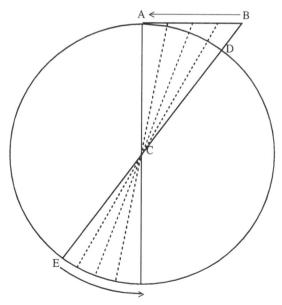

Figure 3.11 As the point D approaches the point A, the line BE approaches the diameter DE and the tangent BA approaches the arc DA.

From [2-B], $DA^2 = (DE)(DB)$. Divide both sides by $ADEA^2$ and obtain $DA^2/(ADEA^2) = (DE)(DB) / (ADEA^2)$. Divide numerator and denominator of the last term by DE and obtain $[DA^2] / [ADEA^2] = [DB] / [(ADEA^2) / (DE)]$, as required here in [2-C]. Invert the expressions and obtain $[ADEA^2] / [DA^2] = [(ADEA^2) / (DE)]/DB$ as required in [2-D] next.

> [2-D] *I obtain the line looked for (namely the third proportional in the ratio of the circumference to the diameter), through which the endeavor of receding from the center would propel a body in the time of one revolution when applied constantly in a straight line.*

"The line [X] looked for" was given in [2-A] by $(X) / (BD) = ADEA^2 / DA^2$. From [2-C], $[ADEA^2] / [DA^2] = [(ADEA^2) / (DE)] / DB$. Substituting this value of $ADEA^2 / DA^2$ into the expression for $(X) / (BD)$, one obtains $(X) / (BD) = [(ADEA^2) / (DE)] / [DB]$, where DE is the diameter of the circle, and $ADEA$ is the circumference (i.e., $ADEA = \pi DE$). Thus, $X = (\pi DE)^2 / DE = \pi^2 DE$, or X is proportional to DE, the diameter of the circle. Newton uses this result in the corollary that follows to demonstrate that the force, which is proportional to the displacement X, is thus proportional to the diameter of the circle, DE.

Step 3. Newton now notes that the force is therefore proportional to the diameter and inversely proportional to the square of the period (i.e., $F \propto X / T^2 \propto DE / T^2$) or what is equivalent, $F = kv^2 / r$.

The text then continues with a numerical calculation of the force of gravity at the equator for an object rotating on the surface of the earth and compares that value to the much larger value given for the "virtue of gravity." The difference between the two values explains why objects do not fly off the rotating earth. Following that discussion, Newton's manuscript contains the following corollary, which relates directly to the functional form of the force necessary to maintain uniform circular motion:

> Corollary. *Hence the endeavors from the centers in diverse circles are as the diameters divided by the squares of the times of revolution*

From the parabolic approximation, the force is inversely proportional to the square of the time and directly proportional to the displacement. Recall that X was the linear distance traveled in a time equal to the period of revolution T under a constant inward radial linear force equal in magnitude to the endeavor of receding from the center. Therefore, the magnitude of the force is inversely proportional to the square of the time T and directly proportional to X and, hence, by [2-D] above, to the diameter DE. Thus, as Newton expresses it, the ratios of "endeavors from the centers in diverse circles as" the ratios $(DE_1 / T_1^2) / (DE_2 / T_2^2)$.

> [Corollary] *or as the diameters multiplied by the [squares of the] number of revolutions made in any same time.*

The number of revolutions N made in any time is proportional to the frequency and hence inversely proportional to the period T (i.e., $T_1 / T_2 = N_2 / N_1$ or $X_1 / X_2 = (DE_1 N_1^2) / (DE_2 N_2^2)$. This proportionality can also be expressed as the square of the tangential speed divided by the diameter (or radius), because the period T is equal to the circumference (and hence π times the diameter D) divided by the tangential speed v. Thus, $D / T^2 = D / (D / v)^2 = v^2 / D$ or the result obtained independently by the Dutch mathematician Christiaan Huygens for the force required to maintain uniform circular motion.[18]

Newton's later dynamics will continue to employ the parabolic approximation used in this early solution, but he will no longer employ the Cartesian terminology of the outward endeavor nor will he see the displacement as an outward radial element. His shift in perspective is dramatic, but it does not require a change in the parabolic approximation. Newton's demonstration of Kepler's law of equal areas will enable him to extend the technique to noncircular and nonuniform orbits; the number and complexity of the mathematical relationships Newton will use to carry solutions to their conclusion will increase; but the underlying parabolic approximation will remain unchanged.

ELLIPTICAL MOTION:
THE CIRCULAR APPROXIMATION

The two techniques for solving direct problems discussed above, the polygonal approximation and the parabolic approximation, do not exhaust the techniques found in Newton's early papers on dynamics. The *Waste Book* also contains a brief statement concerning elliptical motion. In it, Newton states that the force required to maintain elliptical motion can be found from the circle of curvature. In his early work on mathematics, Newton had developed the concept of curvature as a measure of the bending or "crookednesse" of a curve, and in the *Waste Book* he suggested that it could be employed in the analysis of elliptical motion. Newton did not provide any details in this early commentary of how he intended to implement curvature to obtain the force, but it appeared in his later work as an alternate technique for solving direct problems. The following English text is from the *Waste Book* and was written in late 1664 or early 1665.[19]

ELLIPTICAL MOTION

If the body b *moved in an Ellipsis then its force in each point (if its motion in that point bee given) may bee found by a tangent circle of Equall crookednesse with that point of the Ellipsis.*[20]

Among his early original contributions to mathematics, Newton developed a measure for the "crookednesse" or "curvature" of curved figures.

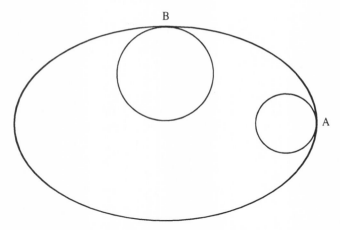

Figure 3.12 Where the ellipse curves most rapidly (*A*) the circle of curvature is small. Where the ellipse curves most slowly (*B*) the circle of curvature is large.

Curvature is a measure of the rate of change of the angle of inclination of the tangent line with respect to the arc length of the curve. In more descriptive terms, the curvature measures the rate of "bending" of the curve at a point. If the curve is a circle, then rate of bending is uniform (i.e., the radius of curvature is everywhere equal to the fixed radius of the circle). If the curve is not a circle, then the rate of bending changes from point to point on the curve. A useful method of measuring the bending of a general curve at a point is to give the curvature of the circle that best approximates the curve at that point, as seen in the ellipse in figure 3.12. Where the ellipse curves most rapidly, the circle of curvature is small and where it curves most slowly, the circle of curvature is large. The circle that represents the best approximation to the curvature—sometimes called the osculating circle—has the same first and second derivatives as the curve at the given point. A circle has constant uniform curvature, an equiangular spiral has curvature that changes uniformly, and conics have curvature that changes systematically but not uniformly. As early as December of 1664 Newton had roughed out a method for finding the center of curvature in an ellipse.[21]

Newton eventually applied curvature as a measure of force and thus applied the dynamics of uniform circular motion to the more complicated problem of elliptical motion. In what I have elected to call the "circular approximation," an elemental arc of a general curve is approximated by an elemental arc of its circle of curvature. Newton employed the circular approximation to solve a number of problems, including that of elliptical motion. The first preserved record of such an analysis occurs in a solution that Newton sent to the British philosopher John Locke in 1690. He may have produced a first draft of that solution as early as 1684, however. This technique also appeared as an alternate solution in the revised editions of the *Principia*. Newton's cryptic statement on elliptical motion, however, indicates that as early as 1665 he had thought of using the circle of curvature and uniform circular motion to analyze motion in an elliptical path. Michael Nauenberg has suggested that Newton may have used this curvature method to obtain numerical and analytical solutions, perhaps as early as 1665. No examples of such calculations have ever been found, however.[22]

CONCLUSION

The *Waste Book* and Newton's other early papers contain many examples of his analysis of problems in dynamics, but the three solutions considered in this chapter are representative of his initial techniques. Before 1669, Newton had set out the three major methods of dynamical analysis that he employed in his more mature dynamics after 1679. First, the polygonal approximation that he developed in 1665 to determine the nature of circular motion in the *Waste Book*, he later used in his demonstration of the generalized nature of Kepler's law of uniform area following 1679. Second, the parabolic approximation that he developed before 1669 in the tract *On Circular Motion*, he later used in his general theorem for solving direct problems in the 1684 tract *On Motion*. Finally, the circular approximation that he alluded to in his statement of late 1664 or early 1665, may have been used to obtain solutions now lost. In any event he employed it in 1690 (or perhaps as early as 1684) as an alternate type of solution for direct problems. All of these techniques were unique to Newton but had their roots in the work of Galileo and Descartes. Newton began with the idealized rectilinear kinematics of Galileo and the terrestrial collisional dynamics of Descartes and from them fashioned three techniques of solving the problems of celestial motions. It is important to remember this dynamic simplicity amid the mathematical complexity of Newton's presentation.

To proceed further with the solution of the Kepler problem of elliptical orbits, one must obtain an expression for noncircular motion that relates the time in orbit to the spatial parameters. The expression that served for the special case of uniform circular motion was obtained from the property of equal angles in equal time and by the relationship between the period and the circumference. Following his correspondence with Robert Hooke in 1679, Newton demonstrated that for any force directed toward a given center the radius from the center to the general orbit sweeps out equal areas in equal times. That demonstration, it has been argued, finally prompted Newton to question and to renounce Decartes's theory of a mechanical celestial vortex as a source of gravitational force. Following his early analysis of uniform circular motion discussed in this chapter, however, Newton appears not to have actively pursued a solution for the direct Kepler problem, other than the reference to the use of curvature. For the next decade, from 1669 to 1679, he was preoccupied with subjects other than dynamics. Thoughts mathematical, optical, theological, and alchemical concerned him, and his interest in the dynamics of the motion of bodies, both terrestrial and celestial, lay fallow until 1679, when he was prompted by correspondence from Robert Hooke to reflect once again upon the nature of the force required to maintain elliptical orbits.

PART TWO

A Guided Study to Newton's Solution

Portrait of Edmund Halley, by Thomas Murray, currently hanging in the Fellows Room of the Royal Society in London. Halley, editor of the first edition of the *Principia*, was elected to the Royal Society in 1678 and served as its clerk from 1686 to 1698 and its secretary from 1717 to 1721. Copyright © The Royal Society. Reproduced by permission.

FOUR

The Paradigm Constructed
On Motion, Theorems 1, 2, and 3

Newton's thoughts on dynamics were awakened late in 1679 by a series of letters from Robert Hooke, who had recently become secretary of the Royal Society and was attempting to revive Newton's interest in contributing to its proceedings. The Royal Society had been founded in London in 1661, and its meetings and publications served to inform the intellectual community of progress in natural philosophy. Newton had been elected to membership in 1672 but had threatened to resign the following year because of criticism of his paper on the theory of colors. Hooke's letter of conciliation of November 1679 opened with the observation that some individuals would misrepresent him to Newton and Newton to him but that differences of opinion, "especially on philosophical issues," should not be the basis of enmity.[1] Hooke offered an olive branch in the form of a request for Newton's opinion on Hooke's hypothesis concerning planetary motion. In contrast to Newton's earlier Cartesian view of curvilinear motion with a tangential endeavor and an outward endeavor counterbalanced by an inward force, Hooke proposed that planetary motion was a dynamic compound of only two motions: a motion along the tangent line due to inertia and a motion toward a central body due to an inward attractive force.[2] The Cartesian outward endeavor is not mentioned.

This revision of the dynamical principle appears not to have been fully appreciated either by Hooke (who never did solve the problem) or by Newton (who did not take up the problem at once). In his immediate reply to Hooke's letter, Newton said that he had not had time to "entertain philosophical meditations" and in fact had "long grudged the time spent in the study."[3] Hooke's initial letter had spoken both of the dynamics of planetary motion and of the astronomer John Flamsteed's claims concerning parallax. Newton had chosen not to respond to the topic of planetary motion but rather spoke to the question of falling bodies on a moving earth, a question that had been raised indirectly by Hooke's report of Flamsteed's measurements of parallax. The correspondence that followed was concerned with that topic. Newton did produce a response to Hooke's planetary challenge, however, at least if one is to trust Newton's recall of his activities some thirty-five years later. In the autumn of 1714, Newton was in the midst of a controversy with the German philosopher and mathematician Wilhelm Gottfried Leibniz over who could claim the invention of calculus, and in the course of that debate, Newton recalled his activities of the winter of 1679.

> In the end of the year 1679 in answer to a Letter from Dr Hook then Secretary of the R.S. I wrote that whereas it had been objected against the diurnal motion of the earth that it would cause bodies to fall to the west, the contrary was true. . . . He had made some experiments thereof and found that they would not fall down to the center of the earth but rise up again and describe an Oval as the Planets do in their orbs. Whereupon I computed what would be the Orb described by the Planets. For I had found before by the sesquialterate proportion of the tempora periodica of the Planets with respect to their distances from the sun, that the forces which kept them in their Orbs about the sun were as the squares of their mean distances from the sun reciprocally: and I found now that whatsoever was the law of the forces which kept the planets in their Orbs, the area described by a Radius drawn from them to the sun would be proportional to the times in which they were described. And by the help of these two Propositions I found that their Orbs would be such Ellipses as Kepler had described.[4]

Newton's statement produces as much confusion as it provides clarification. If taken to be literally true, then he claimed to have solved the inverse problem (given the force, find the orbit) and not the direct problem (given the orbit, find the force). Newton stated that he obtained the knowledge of the inverse square nature of the force from Kepler's third law, in which the cube of the planetary period is proportional to the square of the mean planetary radius. Independent of that result, he determined that the planetary radius sweeps out equal areas in equal times for any central force. Then, using these two propositions, Newton claimed that he found "their Orbs would be such Ellipses as Kepler had described." The exact form of this solution of 1679 must remain a matter of conjecture, however, for no copy of it has ever been found.[5]

Newton may have already settled the direct question for himself in 1679, but the problem of the planets still remained one of the great unanswered questions for the rest of the academic establishment. Halley admitted at a meeting of the Royal Society early in 1684 that he had tried to solve it and failed. Hooke claimed that he had a solution but that he would not reveal it until others tried and realized the difficulty involved. Sir Christopher

Wren offered a small prize of a book to the person who could give him a demonstration within two months, but none was forthcoming within the time period (not even from Hooke). In August of that same year, Halley had occasion to visit Cambridge, perhaps on family business, and took the opportunity to call on Newton, with whom he had become acquainted during discussions of the comet of 1680.[6] In the course of their meeting, the subject of the planetary problem arose. Newton's report of the discussion comes secondhand from the French mathematician Abraham Demoivre who, after Newton's death in 1727, told of a conversation that he had with Newton:

> In 1684 Dr Halley came to visit him [Newton] at Cambridge. . . . The Dr asked him what he thought the Curve would be that would be described by the Planets supposing the force of attraction towards the Sun to be reciprocal to the square of the distance from it. Sir Isaac replied immediately that it would be an Ellipsis. . . . Dr Halley asked him for his calculation without any further delay. Sir Isaac looked among his papers but could not find it, but he promised him to renew it, and then to send it [to] him.[7]

Demoivre recalled that Newton recalled that Halley requested the solution to the inverse problem (i.e., given the force, find the path). In contrast to these recollections, however, stand the copies of the solutions Newton actually produced in fulfillment of his promise to Halley "to renew it [the lost solution of 1679] and then send it to him." The solution that Halley received from Newton after the visit to Cambridge demonstrated that the force, given an elliptical orbit with a force center at a focus, was inversely proportional to the square of the distance (i.e., the direct problem).[8]

True to his promise to Edmund Halley in August of 1684, Newton reproduced a version of the lost solution to the direct Kepler problem that he had generated in 1679 following his correspondence with Robert Hooke. In November of 1684, he sent it to Halley in London. If Halley's initial request in August had been specifically for a solution to the inverse problem, then by December he must have changed his mind, for he received the solution to the direct problem with nothing but exclamations of great admiration.[9] The text that Halley received from Newton, however, contained much more than a determination of the force required to maintain elliptical planetary motion about a focal sun center of force. Newton sent him a tract in which the direct Kepler problem appears as "Problem 3." The tract begins with a set of three definitions, four hypotheses, and two lemmas, each consisting of one or two lines of text. Newton then gave three theorems (with demonstrations), to establish a general procedure for solving direct problems.

> Theorem 1. *All orbiting bodies describe, by radii having been constructed to their center, areas proportional to the times.*

Theorem 2. *For bodies orbiting uniformly on the circumferences of circles, the centripetal forces are as the squares of the arcs described in the same time divided by the radii of the circles.*

Theorem 3. *If a body, by orbiting around the center S, should describe any curved line* APQ, *and if*... [*then*] *I assert that the centripetal force would be reciprocally as the solid* $SP^2 \times QT^2 / QR$, *provided that the quantity of that solid that ultimately occurs when the points* P *and* Q *coalesce is always taken.*

Newton then provided the solutions to three problems as examples of how the general dynamical algorithm of Theorem 3 could be applied to a set of specific direct problems.

Problem 1. *A body orbits on the circumference of a circle; there is required the law of centripetal force being directed to some point on the circumference.*

Problem 2. *A body orbits on a classical ellipse; there is required the law of centripetal force being directed to the center of the ellipse.*

Problem 3. *A body orbits on an ellipse; there is required the law of centripetal force directed to a focus of the ellipse.*

Problem 1 derives the force function necessary to describe uniform circular motion with a force center on the circumference of a circle, and Problem 2 derives the force function necessary to describe elliptical motion with the force center at the center of an ellipse. The solution to Problem 1 is much simpler than the solution to Problem 2, which in turn is simpler than the solution to Problem 3. Further, since neither Problem 1 nor 2 has any obvious physical application, it appears that Newton introduced them as exemplars to demonstrate how Theorem 3 could be used to solve direct problems.[10] The solution to Problem 3, however, does have a physical application. It is the solution to the distinguished direct Kepler problem of the planets (i.e., the force function necessary to describe elliptical motion about a focal force center is an inverse square function of the distance). Although others had anticipated this solution, Newton was the first to produce its demonstration.

The solution to the direct Kepler problem, however, did not exhaust the list of items Newton sent to Halley. Theorem 4 demonstrates Kepler's law relating the periods of planets to their mean radii, and Problem 4 is a version of an inverse problem in which the magnitude of an inverse square force is given and the nature of the resulting orbit is sought. Newton assumes that the body is initially moving in an ellipse with the center of the inverse square force located at a focus of the ellipse, a condition that the solution in Problem 3 of the direct Kepler problem demonstrates is possible (a sufficient condition but not yet demonstrated to be a necessary condition). He then demonstrates that for any other given position and velocity the path will remain an ellipse or will become a parabola or a hyperbola.

> Theorem 4. *Supposing that the centripetal force is reciprocally proportional to the square of the distance from the center, the squares of the periodic times in ellipses are as the cubes of their transverse axes.*
>
> Problem 4. *Supposing that the centripetal force be made reciprocally proportional to the square of the distance from its center, and that the absolute quantity of that force is known; there is required the ellipse which a body will describe when released from a given position with a given speed along a given straight line.*

Newton then turned from his analysis of celestial motion in a nonresistive medium to a comparison of terrestrial motion in nonresistive and resistive media. Problem 5 derives the distance a body would fall freely in a void under an inverse square force, and Problems 6 and 7 combine to extend the analysis to the motion of projectiles in a uniformly resisting medium.

> Problem 5. *Supposing that the centripetal force is reciprocally proportional to the square of the distance from the center, [it is required] to define the spaces which a body falling in a straight line describes in given times.*
>
> Problem 6. *To define the motion of a body carried by its innate force alone through a uniformly resisting medium.*
>
> Problem 7. *Supposing a uniform centripetal force, [it is required] to define the motion of a body ascending and descending straight up and down in a homogeneous medium.*

Newton did not explain why he included these problems of terrestrial motion in a tract that was motivated by questions of celestial motion. It may well be that he wished to call attention to the difference between calculated ideal terrestrial motion in a void and observed terrestrial motion in a resistive medium, such as projectiles fired in the atmosphere of the earth. In contrast to the motion of terrestrial projectiles, the motion of celestial bodies does not display a marked difference between calculated ideal motion in a void and observed celestial motion in the ether. The law of equal areas in equal times is derived for ideal motion in a void, and it appears to correspond to the observed motion in the ether. Thus, any assumption of a mechanical ether, such as that made by Descartes, becomes suspect. If the ether is so fine as to offer no resistance to the motion of celestial bodies, then it surely is too fine to provide the necessary mechanical collisions to account for the gravitational force. The implication of such a difference concerning the role of the ether in producing the gravitational force has been discussed in chapter 2.[11]

The tract that Halley received from Newton in November of 1684 contained only a fraction of what would shortly be demonstrated in the

1687 edition of the *Principia*, but it was much more than anyone had thus far been able to accomplish, and much more than Halley had requested. Shortly after receiving it, Halley made yet another journey to Cambridge in order to consult Newton once again. He found Newton already at work revising and enlarging the tract into what would ultimately become the 1687 edition of the *Principia*. Halley returned to London and early in December 1684 he communicated the initial copy of the tract to the Royal Society in order to secure Newton's rights of authorship, even as work on the extended version continued. What follows includes a translation of the Latin copy of the tract that appears in the *Register Book* of the Royal Society and now bears the title, *On Motion of Bodies in Orbit* or, simply, *On Motion*. The version in the *Register Book* contains some obvious mistakes of transcription, particularly in the diagrams. Corrections were made to the transcript used in this translation by comparing it to other versions of this tract. Although the authenticity of the text is not in question, the original manuscript that Newton sent to Halley has never been found.[12]

INTRODUCTION:
DEFINITIONS, HYPOTHESES, AND LEMMAS

Definitions

Newton began *On Motion* with three definitions: centripetal force, innate force, and resistance. The list of definitions and the details of their descriptions were greatly enlarged in subsequent versions of the work, but in this first text they were set out very simply and compactly.

> Definition 1. *I call centripetal the force by which a body is impelled or attracted toward some point which is regarded as the center.*

Definition 1 contains the first use of the term "centripetal" (center-seeking), which Newton coined as a complement to the term "centrifugal" (center-fleeing) that Huygens had employed in his writings.[13] The term signaled a major clarification of Newton's analysis of dynamics as found in the demonstrations of circular motion in the *Waste Book* (1665) and *On Circular Motion* (pre-1669) that were discussed in chapter 3. In those works, Newton, consistent with Descartes, refers to an outward endeavor, and in *On Motion* Newton, consistent with Hooke, does not mention an outward endeavor. Newton's rejection of both Cartesian perspective and terminology is not a change in Newton's method of demonstration, however. In both the early work and this later tract Newton employs the parabolic approximation in which the force is directly proportional to the radial displacement and inversely proportional to the square of the time. Moreover, Newton continued to employ the term "centrifugal" in other contexts, well after the writing of this work.[14]

Definition 2. *Moreover, [I call] the force of a body, [the force] innate in a body, that by which it endeavors to persevere in its own motion along a straight line.*

In Newton's analysis, the use of the "force innate in a body" most closely conforms to the contemporary use of the "magnitude of the linear momentum." Therefore, Newton's use of the word "force" in this context is at variance with modern usage, which reserves the term "force" for the "time rate of change of the linear momentum." Regardless of his choice of term, he uses "innate force" in a manner consistent with modern analysis.

Definition 3. *And [I call] resistance the force which comes from a regularly impeding medium.*

The topic of motion in a resistive medium will not be included in the material covered in this study.

Hypotheses

Newton provided four hypotheses: the first sets out assumptions concerning resistance to motion; the second describes force-free motion; the third states the parallelogram rule for the addition of displacements produced by separate forces; and the fourth expresses a version of Galileo's time-squared dependence of linear displacement under a constant force.

Hypothesis 1. *In the next nine propositions the resistance is zero; in those propositions following, the resistance is conjointly as the speed of the body and the density of the medium.*

As noted, the first nine propositions consist of four theorems and five problems, all of which are concerned with ideal motion in the absence of resistance. The final two propositions (Problems 5 and 6) treat motion in a resistive medium.

Hypothesis 2. *Every body by its innate force alone progresses uniformly along a straight line to infinity unless something impedes it from outside.*

Following Descartes, Newton states that motion free from an external force (i.e., motion subject only to "innate force") takes place at a uniform rate along an infinite straight line. An enlarged version of this statement appears as the first law of motion in the *Principia*.[15]

Hypothesis 3. *A body, in a given time, with forces having been conjoined, is carried to the place where it is carried by separated forces in successively equal times.*

This rule for the combination of displacements as a measure of forces was implicit in Newton's pre-1665 analysis of circular motion but is made explicit here. Although Newton gives it as a hypothesis in *On Motion*, in the first revision of the tract Newton adds a demonstration and promotes the hypothesis to the status of a lemma.[16]

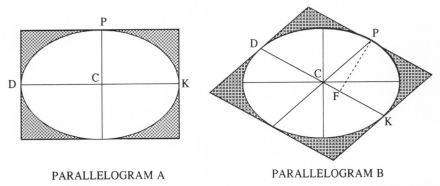

PARALLELOGRAM A   PARALLELOGRAM B

Figure 4.1 The area of parallelogram A is equal to the area of parallelogram B (Proposition 31, Book 7, of the *Conics* of Apollonius of Perga).

Hypothesis 4. *The space which a body, with some centripetal force impelling it, describes at the very beginning of its motion, is in the doubled ratio of the time.*

This relationship is critical to all of Newton's analysis of action under a continuous centripetal force. It is the very core of his analysis; yet it is given here very simply and with little explanation. This hypothesis will be revised by the addition of a demonstration, and it also will be promoted to the status of a lemma.

Lemmas

Lemma 1. *All parallelograms described around a given ellipse are equal to each other. This is established from the* Conics.

This lemma is demonstrated in Book 7, Proposition 31 in the *Conics* of Apollonius of Perga (c. 262–c. 200 B.C.); see figure 4.1.[17] The area of the circumscribed parallelogram *A* is 2*PC* x *DK*, and it is equal to the area of the circumscribed parallelogram *B*, which is 2*PF* x *DK*. It is important to note in parallelogram *B* that *PF* is the normal to *DK*, while in parallelogram *A* that *PC* is the normal to *DK*. The sides of parallelogram *B* are tangent to the ellipse at points *P* and *D*, where *DK* is constructed parallel to the line tangent at point *P*. This relationship will appear as Lemma 12 in the *Principia*, where it is employed in the solution of the direct Kepler problem.

Lemma 2. *Quantities proportional to their differences are continuously proportional.* Set A: (A − B) = B: (B − C) = C: (C − D) = . . . *and by dividing there will be produced* A: B = B: C = C: D = . . .

This lemma has application only to motion in a resistive medium, a topic which does not appear in the first three sections of Book One of the *Principia*.

Following the completion of the draft of *On Motion of Bodies in Orbit* (or simply *On Motion*) that was sent to Halley, and hence to the Royal Society, Newton produced a slightly enlarged version of the tract, entitled *On the Motion of Spherical Bodies in Fluids*.[18] The body of the tract, which supplied the method and solutions to the direct problem in nonresisting media, was essentially unchanged from the first draft (except for the addition of a paragraph at the end of the scholium to Theorem 4). Newton did expand, however, the rather sparse statement of the fundamental hypotheses that was just discussed.

Hypothesis 3 Becomes Lemma 1

Hypothesis 3. *A body, in a given time, with forces having been conjoined, is carried to the place where it is carried by separated forces in successively equal times.*

Hypothesis 3 of the first draft of *On Motion* now appears as Lemma 1 in the second draft. The initial statement is slightly revised, and Newton appends a detailed demonstration. An earlier version of this parallelogram rule was discussed in chapter 3 during the analysis of uniform circular motion. It is important to note that the measure of a force is the displacement it produces in a given time, and it is the displacements that are combined when the "forces are conjoined."

Lemma 1. *A body, with forces having been conjoined, describes the diagonal of a parallelogram in the same time as it describes the sides, with [forces] having been separated.*

If a body in a given time were to be carried from A to B by the action of the force M alone and from A to C by the force N alone, [then] complete the parallelogram ABDC, and it will be carried in the same time from A to D by both forces. [See fig. 4.2.]

For since force N acts along the line AC parallel to BD, by Law 2 this force [N] will do nothing to change the speed of [the body's] approaching the line BD, impressed by the other force [M]. The body will therefore approach the line BD in the same time whether the force AC [N] is impressed or not; and so at the end of that time it will be found somewhere on the line BD. By the same reasoning it will at the end of the same time be found somewhere on the line CD, and consequently must be found at the meeting D of both lines.

Newton applied the parallelogram rule implicitly in all his dynamics. He does not, however, give an explicit formal defense of the application of this lemma to the polygonal and parabolic approximations, either in this tract or in the first edition of the *Principia*. Only in the revised editions of the *Principia* does he offer an explicit demonstration of its application.[19]

Figure 4.3 displays the situation to which the parallelogram rule is applied in the format of the polygonal approximation in Theorem 1 (to follow). A body moves with uniform rectilinear motion from point *A* to point

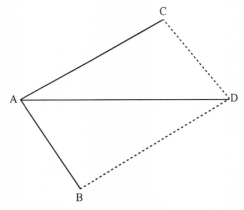

Figure 4.2 Based on Newton's drawing for the demonstration of Lemma 1.

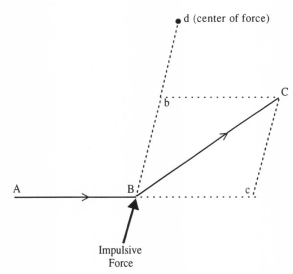

Figure 4.3 The parallelogram rule as applied to the polygonal approximation. In a given time, *Bc* is the displacement due to the initial "innate force," *Bb* is the displacement due to the impulsive force, and *BC* is the resultant displacement.

B in a given time ΔT. At point B an impulsive centripetal force, directed toward the center of force at point d, acts for a vanishingly small time δt on the body. The body then moves with uniform rectilinear motion from point B to point C in the given time ΔT. If the impulsive force had not acted at point B, then the body would have moved with uniform rectilinear motion from point B to point c in the time ΔT under the action of only its initial motion. If the body had been at rest at point B, then it would have moved with uniform rectilinear motion from point B to point b in the time ΔT under the action of only the motion produced by the impulsive force. (Note that the displacement Bb must be in the direction of the line of action of the impulsive centripetal force.) The composite uniform motion from point B to point C is along the diagonal of the parallelogram formed by the initial motion Bc (or bC) and the added motion Bb (or cC). Thus, a body is carried in a given time $[\Delta T]$ by combined forces [the initial "force of the body's motion" plus the "change in the body's motion" due to the impulse] to the place $[B \rightarrow C]$ where it is carried by separated forces [$B \rightarrow c$ by the initial motion and $B \rightarrow b$ by the impulsive motion] in successively equal times $[\Delta T]$.

If the force is continuous rather than impulsive, then the curve is also continuous rather than polygonal. For a continuous force the body moves along the curved path between the points B and C, and Newton considers a situation in which the point C approaches very closely to the point B. Thus, the interval of time is extremely small, and the force can be assumed to be constant over that interval. Galileo has demonstrated that ideal projectile motion under a constant force is parabolic, and hence the element of arc BC is approximated by a parabolic element (i.e., the parabolic approximation discussed in chapters 1 and 2). The displacement BC of any future point C on the elemental parabolic arc can be found by using the parallelogram rule to combine the displacement BC due to the initial tangential velocity with the deviation Bb due to the constant force. Thus, this rule is applied to both impulsive and continuous forces in a consistent manner.

THEOREM 1: THE LAW OF EQUAL AREAS IN EQUAL TIMES

In Newton's theorems that follow Theorem 1, the path of the particle appears as a continuous curve and the force changes continuously as the particle traverses the path. In Theorem 1, however, the particle travels with a given uniform rectilinear motion between points A and B. At point B it experiences an impulsive force and then travels with a different uniform rectilinear motion between points B and C. At point C it experiences

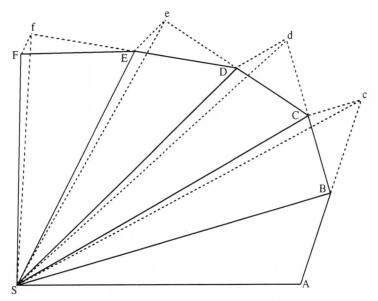

Figure 4.4 Based on Newton's drawing for Theorem 1.

yet another impulsive force. The process is repeated again and again. Ultimately, Newton required that the distance between points become infinitely small, and thus he required the polygonal path of the particle to become a smooth curve. This type of approximation is similar to the polygonal approximation that Newton used in his early work of 1665 on uniform circular motion and is particularly well suited to the demonstration of this theorem. Figure 4.4 is based on Newton's diagram that accompanies Theorem 1 in *On Motion*; it is instructive to note how the diagram is constructed.

> Line *AB*. One starts at point *A* and draws a line of arbitrary length to point *B*. The line *AB* represents the displacement of the body for a given time with a given velocity. At point *B* an impulsive force that is directed toward the center *S* acts on the body.
> Line *Bc*. Then the line is extended from point *B* to point *c*, where line *Bc* is equal in length to line *AB*. The line *Bc* represents the displacement that would have taken place in the given interval of time if no force had acted at point *B*.
> Line *BC*. Because an impulsive force does act at point *B*, however, one must construct a line *BC* that represents the actual displacement that does take place in that given time.
> Line *Cc*. Then the actual displacement *BC* is connected to the hypothet-

ical force-free displacement *Bc* by a line *Cc*. The line *Cc* represents the deviation of the body from *Bc* due to the action of the impulsive force at *B*. Therefore, *Cc* is constructed parallel to the line of force *SB* because the displacement *Cc* must be in the same direction as the line of action of the impulsive force. The rest of the construction is a repetition of the preceding procedure.

It simply remains to demonstrate (1) that the area *ASB* is equal to the area *BSc* and (2) that area *BSc* is equal to the area *BSC*. Then, in a given time, (3) area *ASB* is equal to the area *BSC*, which in turn is equal to area *CSD*, and so on. Thus, for a series of discrete impulsive forces, equal areas will be swept out in equal times. The correspondence with a force that acts continuously is achieved by passing to the limit by letting "the triangles be infinite in number and infinitely small, so that the individual triangles correspond to the individual moments of time."

What follows is Newton's statement of this important theorem, one line at a time, each line followed by a detailed discussion.

Demonstration

Theorem 1. *All orbiting bodies describe, by radii having been constructed to their center, areas proportional to the times.*

The only restriction on the force is that it be directed to a given point *S*. For such a central force, Newton demonstrates that the radius linking the point *S* and the body *P* sweeps out equal areas in equal times. For the special example of uniform circular motion about a point at the center of the circle, the radius also sweeps out equal angles in equal times. In his early analysis of such motion, Newton employed this more restricted angular version of the area law to relate the period and circumference to arc and deviation, and thus to express the results of the time/distance relationship in the desired form (see chapter 2). In Theorem 1, however, Newton demonstrates that the law holds true for any centripetal force acting in a medium devoid of resistance, and thus he can express the time, and hence the force, in terms of the geometric elements of the figure for nonuniform motion.

[A] *Let the time be divided into equal parts, and in the first part of the time let a body by its innate force describe the straight line* AB.

The line *AB* is the displacement due to the uniform rectilinear motion of the body between points *A* and *B*. The body leaves point *A* with a given speed and travels to point *B* in the absence of any external force.

[B] *The same body would then, if nothing impeded it, proceed directly to* c *in the second part of the time, describing the line* Bc *equal to itself* AB,

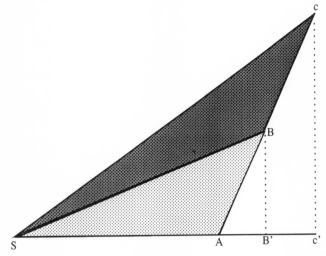

Figure 4.5 The area of triangle *ASc* is equal to twice the area of triangle *ASB* and thus area *ASB* equals area *BSc*.

In this theorem rectilinear motion along the tangent occurs when nothing impedes the tangential motion (i.e., in the *absence of a centripetal* "center-seeking" force). In the pre-1669 tract, *On Circular Motion*, however, rectilinear motion along the tangent occurs when there is no impediment to the Cartesian outward endeavor (i.e., in the *presence of a centrifugal* "center-fleeing" endeavor). After 1679, however, Newton set aside the Cartesian outward endeavor and was concerned only with the two dynamical elements that Hooke stressed in his early paper, the tangential displacement and the center-seeking force.

[C] *so that, when the radii* AS, BS, *and* cS *were extended to the center, areas* ASB *and* BSc *would be made equal.*

See figure 4.5. The line segment *AB* is the displacement the body makes during the first time interval and the line segment *Bc* is the displacement, equal to *AB*, that the body would have made in an equal time if no impulsive force had acted upon it at point *B*. Thus, the displacement *Ac* is equal to twice the displacement *AB* and thus the height *cc'* is equal to twice the height *BB'*. The area of a triangle is equal to one-half the product of the base and the height. Triangles *ASB* and *ASc* have a common base *SA* and the height of triangle *ASc* is twice the area of triangle *ASB*. Thus, the area of triangle *ASc* is twice the area of triangle *ASB*. From figure 4.5 one can express the area of triangle *BSc* as the difference between the area of triangle *ASc* and the area of triangle *ASB*. Thus, area *BSc* = area *ASc* − area *ASB* and, because area *ASc* = 2 area *ASB*, then area *BSc* = area *ASB*.

THE PARADIGM CONSTRUCTED 83

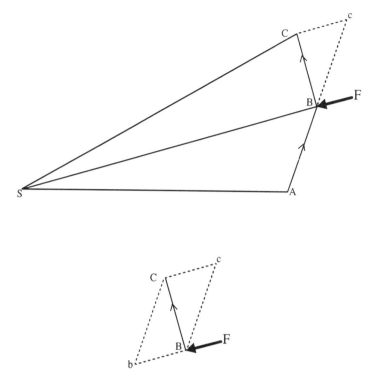

Figure 4.6 The parallelogram rule as applied to Theorem 1.

[D] *Now when the body comes to B, let the centripetal force act with one great impulse, and let it make the body deflect from the straight line Bc and proceed along the straight line BC.*

In the early discussion (1665) of circular motion in the *Waste Book*, the "force or pression" came from the "collision" or "reflection" of the ball from the curve and was the impediment to the "outward endeavor." In this more mature work (1684), the impulse delivered to the ball comes from an external unbalanced "centripetal force" and there is no mention of the Cartesian "outward endeavor."

[E] *Parallel to the same BS, let cC be extended, meeting BC at C, and when the second interval of time is finished, the body will be found at C.*

See figure 4.6. If no force had acted at point *B*, then the particle would have made the hypothetical displacement *Bc* equal in length to the initial displacement *AB*. Because a force *F* does act at point *B*, however, the particle is diverted and instead makes the actual displacement *BC*. The deviation or difference between the hypothetical displacement *Bc* and the

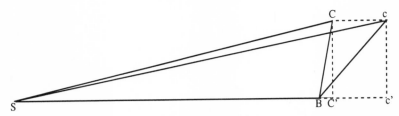

Figure 4.7 The area of triangle *SBC* equals the area of triangle *SBc*.

actual displacement *BC* is given by the dashed line *Cc*. The small construction in figure 4.6 is an application of the parallelogram rule given in Hypothesis 3. The two uniform rectilinear motions to be combined at point *B* are (1) the uniform motion along *Bc*, which was retained from the body's initial motion at point *B*, and (2) the uniform motion along *Bb*, which was generated by the impulsive force *F* at point *b*. The resultant uniform rectilinear motion is along *BC*. In the given time, the particle will make the displacement *BC*, where *BC* is the diagonal of the parallelogram *BbCc* formed with sides *Bc* and *Bb*, where *Bc* and *Bb* are the displacements the body would have made separately in the given time. Consistent with the parallelogram rule, the deviation *Cc* is constructed parallel to the line *Bb* and hence parallel to the line of force *SB*.

> [F] *Join* S *and* C *and because of the parallels* SB *and* Cc, *the triangle* SCB *will be equal to the triangle* SBc *and hence also to the triangle* SAB.

See figure 4.7. The triangles *SBC* and *SBc* have the same base *SB* and, because the deviation *Cc* is parallel to the line of force *SB*, they have the same perpendicular to that base through *C* and *c* (i.e., the height *C'C* equals the height *c'c*). Thus, the areas of triangles *SBC* and *SBc* are equal because they have a common base and equal heights.

> [G] *By a similar argument, if the centripetal force should act successively at* C, D, E, *etc., making the body in separate moments of time describe the separate straight lines* CD, DE, EF, *etc., the triangle* SCD *will be equal to the triangle* SBC, SDE *to* SCD, SEF *to* SDE (*and so on*).

Simply repeat [F] for each of the successive blows at the points *C, D, E, . . .* , that are shown above in the original drawing, figure 4.4.

> [H] *In equal times, therefore, equal areas are described.*

The conclusion is drawn from [G].

> [I] *Now let these triangles be infinite in number and infinitely small, so that the individual triangles correspond to the individual moments of time, the centripetal force acting without interruption, and the proposition will be established.*

THE PARADIGM CONSTRUCTED

This limiting procedure is similar to that which Newton employed in the polygonal approximation solution to the problem of uniform circular motion in the early work in the *Waste Book* (discussed in chapter 3). Theorem 1 above becomes Proposition 1 in the 1687 edition of the *Principia* much in the form that it appears here, and Newton revises and extends it in the later editions.[20]

Newton's demonstration of Kepler's law of equal areas in equal times in Theorem 1 is a major step forward in his construction of a paradigm for the solution of direct problems. The challenge of the direct problem is to find the functional dependence of the force upon the distance between the body and the center of force necessary to describe the orbital motion. The only requirement imposed on the nature of the force by Theorem 1 is that it be directed toward a fixed center. Theorem 1 enables Newton to express the time of motion in terms of the area swept out and hence in terms of the dimensions of the orbit. He employs it and the parabolic approximation in Theorem 3 to develop a general measure of centripetal force.

THEOREM 2: UNIFORM CIRCULAR MOTION

Theorem 2 on uniform circular motion employs the parabolic approximation as in the pre-1669 tract *On Circular Motion* rather than the polygonal approximation as found in the solution given in the *Waste Book* (both solutions are discussed in chapter 2). In Corollary 2 of the pre-1669 tract *On Circular Motion*, Newton wrote that "the endeavors from the centers in diverse circles are as the diameters divided by the squares of the times of revolution." Here in Corollary 2 of the post-1679 tract he wrote that "[the centripetal forces are] reciprocally as the squares of the periodic times divided by the radii." The change from diameters to radii is trivial, but the shift in emphasis from the "outward endeavor" to the center-seeking "centripetal force" is significant. For uniform circular motion, the arcs are proportional to the speeds. Newton elected to express the theorem in terms of the arc, however, and to reserve the statement in terms of the speeds for the first corollary, where the force is given as proportional to the square of the speeds divided by the radii.[21]

Figure 4.8 is based on Newton's diagram that accompanies this theorem in *On Motion*. There are two circles because the results are to be expressed as the ratio of the forces required to maintain uniform circular motion for two different radii, *SB* and *sb*. As in his earlier analysis, the line segments *BC* and *bc* represent the tangential displacements that would have taken place if no force had acted on the bodies at points *B* and *b*; the arcs *DB* and *db* are the actual circular paths followed under the action of the centripetal forces; and the deviations that measure the forces are the

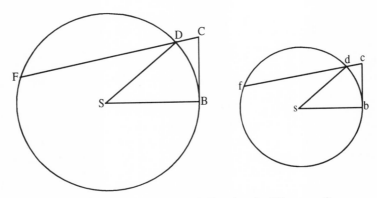

Figure 4.8 Based on Newton's drawing for Theorem 2.

line segments *DC* and *dc*. Newton's shift in emphasis from a Cartesian outward endeavor to an inward centripetal force does not affect his use of the parabolic approximation. In the limit, the force still is assumed to be approximately constant and its magnitude still is proportional to the deviation *DB* and inversely proportional to the square of the time. The following lines from the text are interspersed with detailed explanations and discussions.

Demonstration

[A] Theorem 2. *Let the bodies* B *and* b *orbiting on the circumferences of the circles* BD *and* bd *describe in the same time the arcs* BD *and* bd. *By their innate force alone they would describe the tangent lines* BC *and* bc *equal to these arcs.*

As in previous examples, the important elements shown in the diagram in figure 4.8 above are the inertial displacements *BC* and *bc*, which would have taken place along the tangent lines in the absence of a force, and the arc lengths *BD* and *bd* of the actual path. The arc lengths *BD* or *bd* in a given time are equal in length to the tangent lengths *BC* or *bc*. If the tangent *BC* were wrapped around the circle, then the end of the tangent *C* would travel on a curved arc *CD* (its involute) about *B* terminating on point *D* (see fig. 4.9).[22] Thus, the tangent length *BC* is exactly equal to the arc length *BD*.

[B] *The centripetal forces are those that perpetually draw bodies back from the tangents toward the circumferences* [*of the circles*], *and hence are to each other as the distances* CD *and* cd *surmounted by them,*

Or, $F_C / f_c = CD / cd$, where F_C and f_c are the centripetal forces and *CD* and *cd* are the deviations from the tangential paths relative to the circular

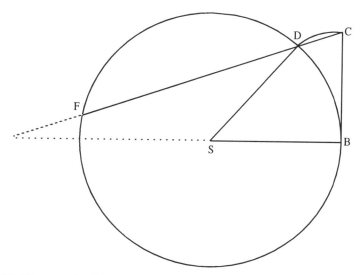

Figure 4.9 The length of the arc *BD* is equal to the length of the tangent segment *BC*.

paths. The deviation *CD* is the chord of the arc *CD* about *B* (see fig. 4.9) and the deviation *cd* is the chord of the arc *cd* about *b*. In his pre-1669 analysis (see chapter 3) Newton spoke of the outward endeavor, and the deviation *CD* was directed incorrectly along a diameter of the circle, as suggested by the Cartesian terminology. Figure 4.10 is a comparison of the form of the diagram in his pre-1669 analysis of uniform circular motion with the form of the diagram in this post-1679 analysis. Note how in this later figure the deviation *CD* is not along a diameter but along the chord *DF*. This change in the slope and location of *CD* is now determined by the correct requirement in [A] that the tangent length *BC* must equal the arc length *BD* (see fig. 4.9). The deviation *CD* is no longer a potential but impeded outward displacement that acts in a radial direction, as suggested by Descartes. The revision of the diagram does not invalidate Newton's earlier solution because the Euclidean theorem to be employed in the next step is valid for both chords and diameters. The forces in either case are proportional to the deviations *DC* and *dc*, or $F_C / f_c = CD / cd$, as required.

[C] *that is, on producing* CD *to* F *and* cd *to* f, *as* BC² / CF *to* bc² / cf *or as* BD² / (1/2)CF *to* bd² / (1/2)cf.

Or $F_C / f_c = CD / cd = (BC^2 / CF) / (bc^2 / cf) = (BC^2 / (1/2)CF) / (bc^2 / (1/2)cf)$. In the earlier pre-1669 drawing (see fig. 4.10), Newton specifically called upon Proposition 36 of Book 3 of Euclid's *Elements* to relate the lines *CB* (the tangent), *CD* (the deviation), and *CE* (the diameter plus the

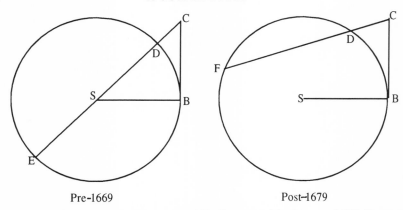

Figure 4.10 A comparison of the pre-1669 diagram with the post-1679 diagram in the analysis of circular motion.

deviation) as follows: $CE / CB = CB / CD$. In this later text he did not give a specific reference to Euclid but the same relationship holds for the lines CB, CD, and CF (the chord of the circle plus the deviation) in the revised post-1679 diagram. In figure 4.11A, Euclid's theorem is valid when CF contains *any* chord of the circle.[23] Thus, in figure 4.11B, the deviation CD (dashed line) is obtained from $CD / BC = BC / CF$ or $CD = BC^2 / CF$. Newton chose to express the results in terms of the radius $(\frac{1}{2}) CF$ rather than the diameter CF. Thus, $F / f = CD / cd = (BC^2 / CF) / (bc^2 / cf) = (BC^2 / (\frac{1}{2}) CF) / (bc^2 / (\frac{1}{2}) cf)$.

> [D] *I am speaking of the very minute distances* BD *and* bd, *to be diminished into infinity, so that in place of* $(\frac{1}{2})$CF *and* $(\frac{1}{2})$cf, *it would be allowed to write the radii* SB *and* sb *of the circles. This done, the Proposition will be established.*

Or, as point D approaches point B, then line CF approaches line $2SB$ (see fig. 4.12). And, $(\frac{1}{2}) CF$ can be replaced by the radius SB. Thus, the proposition is demonstrated, and the forces are proportional to the square of their arcs divided by the radii of their circles (i.e., $F_C / f_c = (BC^2 / ((\frac{1}{2}) CF) / (bc^2 / ((\frac{1}{2}) cf) = (BC^2 / SB) / (bc^2 / (sb))$.

> Corollary 1. [*Hence*] *the centripetal forces are as the squares of the speeds divided by the radii of the circles.*

For uniform circular motion, the arc is proportional to the tangential speed. From [D], the ratios of the forces $F_1 / F_2 = (arc_1^2 / r_1) / (arc_2^2 / r_2)$ and because $arc_1 / arc_2 = v_1 / v_2$ then $F_1 / F_2 = (v_1^2 / r_1) / (v_2^2 / r_2)$.

> Corollary 2. *And reciprocally as the squares of the periodic times divided by the radii of the circles.*

Or, $F_1 / F_2 = (T_2^2 / r_2) / (T_1^2 / r_1)$. The tangential speed v = circumference/period = $2\pi r / T$, thus, $v^2/r = 4\pi^2 (r/T)^2 / r$. From Corollary 1, $F_1 /$

THE PARADIGM CONSTRUCTED

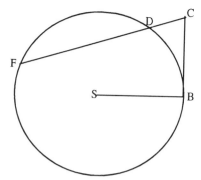

Figure 4.11A From Proposition 36 of Book 3 of Euclid's *Elements*, one has $CD/BC = BC/CF$.

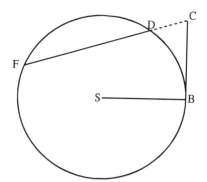

Figure 4.11B In the diagram for Theorem 2, one has $CD/BC = BC/CF$ or the deviation $CD = BC^2/CF$.

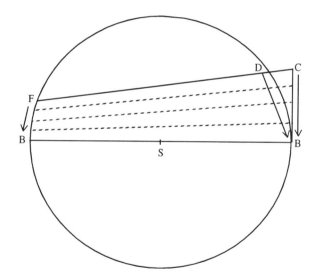

Figure 4.12 As the point D approaches the point B, the line CF approaches the diameter $2SB$.

$F_2 = (v_1^2/r_1)/(v_2^2/r_2)$, which becomes $(4\pi^2(r_1/T_1)^2/r_1)/(4\pi^2(r_2/T_2)^2/r_2) = (r_1/T_1^2)/(r_2/T_2^2) = (T_2^2/r_2)/(T_1^2/r_1)$, or the force is reciprocally proportional to the squares of the periods divided by the radii.

> Corollary 3. *From this, if the squares of the periodic times are as the radii of the circles, [then] the centripetal forces are equal, and conversely.*

If T^2 is proportional to r, then $(T_2^2 / r_2) / (T_1^2 / r_1) = 1$. From Corollary 2, $F_1 / F_2 = (T_2^2 / r_2) / (T_1^2 / r_1) = 1$ and thus $F_1 = F_2$.

Corollary 4. *If the squares of the periodic times are as the squares of the radii, [then] the centripetal forces are reciprocally as the radii, and conversely.*

Or, $F_1 / F_2 = r_2 / r_1$. If T^2 is proportional to r^2, then from Corollary 2, $(T_2^2 / r_1) / (T_1^2 / r_2) = r_2 / r_1$ and thus $F_1 / F_2 = r_2 / r_1$.

Corollary 5. *If the squares of the periodic times are as the cubes of the radii, [then] the centripetal forces are reciprocally as the squares of the radii, and conversely.*

Or, $F_1 / F_2 = r_2^2 / r_1^2$. If T^2 is proportional to r^3, then from Corollary 2, $(T_2^2 / r_2) / (T_1^2 / r_1) = r_2^2 / r_1^2$ and thus $F_1 / F_2 = r_2^2 / r_1^2$.

Scholium

The case of the fifth corollary holds true in the celestial bodies. The squares of the periodic times are as the cubes of the distances from the common center around which they revolve. Astronomers already agree that this holds true in the major planets orbiting around the sun and in the minor ones around Jupiter and Saturn.

Newton did not give credit here to Kepler for the relationship in this scholium, but the result is found in Book 5 of Kepler's *World Harmony* (1619).[24] In Theorem 4 of this tract, Newton demonstrates that for elliptical motion the square of the period is as the cube of the transverse (major) axis of the ellipse.

THEOREM 3: THE LINEAR DYNAMICS RATIO

Theorem 3 is Newton's crowning achievement in dynamics. It sets forth the combination of geometric elements that provide the basic paradigm for his primary solution of direct problems. Figure 4.13 makes visual these basic geometric elements: the displacement *QR* is the deviation from the linear path *PR* produced by the centripetal force directed toward the center of force *S*, and the triangular area *PSQ* is given by one-half of the product *QT* x *SP*. In the limit as the point *Q* approaches the point *P*, the force is assumed to be constant and its magnitude is proportional to the displacement *QR* divided by the square of the time (i.e., the parabolic approximation). From Theorem 1, the time is proportional to the area *PSQ*. Thus, combining the parabolic approximation and the area law, the force is shown to be proportional to the linear dynamics ratio $QR / (QT^2 \times SP^2)$, or as Newton preferred, "the centripetal force would be reciprocally as the solid $(SP^2 \times QT^2) / QR$."

The results of Theorem 3 are valid for any general curve and are not restricted to circles or ellipses (for example, Newton employed it to calculate the force necessary to produce a spiral motion). Nevertheless, New-

THE PARADIGM CONSTRUCTED 91

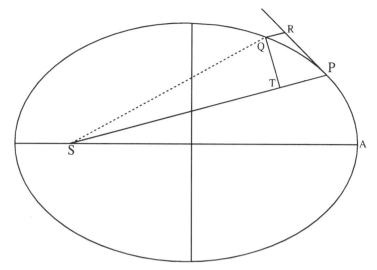

Figure 4.13 Based on Newton's diagram for Theorem 3.

ton's thoughts must have been on the distinguished Kepler problem when he drew the diagram for this theorem, because the general curve in figure 4.13 looks very much like an ellipse (in fact, it is an ellipse). He does not employ any of the specific properties of an ellipse, however, in obtaining the general relationship between the force and the dynamic elements of the general curve. In the following, the theorem is divided into a number of separate lines and detailed commentaries are given for each line.

Demonstration

[A] Theorem 3. *In the indefinitely small figure QRPT the line segment QR is, with the time given, as the centripetal force,*

Or, $QR \propto F$ for a given time. The applied force is assumed to be approximately constant in magnitude and direction in the limit as the point Q approaches the point P, or as Newton puts it, "in the indefinitely small figure *QRPT.*" Thus, the elemental arc of the general curve can be approximated by an elemental arc of a parabola. The tangential displacement *PR* is the inertial displacement the body would have made in a given time in the absence of a centripetal force under the sole action of the velocity it had at point P. The deviation QR is the linear displacement the body would have made in the same given time under the sole action of a constant applied force directed along the line *SP* from the initial point P to the center of force S. The resultant displacement PQ is given from the parallelogram rule by the sum of the tangential displacement PR due to the initial "innate force" at point P and the linear displacement QR due to the continuous applied force. [In terms of modern notation, the displacement $QR = (1/2)(F/m)(t^2)$ for a constant force F and mass m, where F/m is the acceleration. Thus, for a given time, $QR = (t^2/2m)F = k_1 F$, (where k_1 is a constant equal to $t^2/2m$) or the displacement QR is directly proportional to the force F.]

[B] *and, with the force given, as the square of the time,*

Or, $QR \propto t^2$ for a given force. This condition follows from Hypothesis 4 (or in the revised Lemma 2: "The space with which a body, with some centripetal force impelling it, describes at the very beginning of its motion, is in the doubled ratio of the time"). [In modern notation, from section [A], $QR = (1/2)(F/m)(t^2) = (F/2m)(t^2) = k_2 t^2$, where k_2 is a constant equal to $F/2m$. Thus, for a given force F the displacement QR is directly proportional to the square of the time t.]

[C] *and hence, when neither is given, as the centripetal force and the square of the time conjointly;*

Or, $QR \propto (F)(t^2)$. From the expressions [A] and [B], when neither the time nor the force is given, the line segment QR depends directly upon both the force and the square of the time. [In modern notation, $QR = (1/2m)(F)(t^2) = k_3 (F)(t^2)$, where k_3 is a constant equal to $1/2m$. Thus, QR is directly proportional to the product of the force and the square of the time.]

[D] *that is, as the centripetal force taken once and the area* SQP *proportional to the time (or its double,* SP x QT) *taken twice.*

Or, $QR \propto (F)(t^2) \propto (F)(SP \times QT)^2$. From [C], the displacement is proportional to the first power of the force ("taken once") and the second power of the time ("taken twice"). From Theorem 1, the time is proportional to the triangular area SQP, which is equal to one-half the base SP times the height QT (see fig. 4.14), or the product $SP \times QT$ is the double of the area where ultimately the arc QP will approach the chord QP.

[E] *Let each part of this proportionality be divided by the line segment* QR *and there will result unity as the centripetal force and* SP² x QT² / QR *conjointly, that is the centripetal force reciprocally as* SP² x QT² / QR. *Which was to be proven.*

Or, $1/F \propto (SP^2 \times QT^2)/(QR)$. Thus, "the centripetal force would be reciprocally as the solid $SP^2 \times QT^2 / QR$." [Divide $QR = k_3(F)(SP^2 \times QT^2)$ by QR, and one has $1 = k_3(F)(SP^2 \times QT^2)/QR = k_3(F)/(QR)/(SP^2 \times QT^2)$, or $1/F = k_3(SP^2 \times QT^2)/(QR)$.]

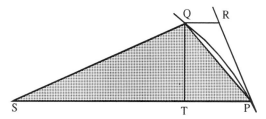

Figure 4.14 The shaded triangular area is equal to one-half of the product of the base *SP* and the height *QT*. As the point *Q* approaches the point *P*, the arc *QP* approaches the chord *QP*.

Corollary. *Hence if any figure is given, and on it a point toward which the centripetal force is directed, [then] it is possible for a law of centripetal force to be found which will make a body orbit on the perimeter of that figure.*

The result above represents a general theorem to be employed in obtaining solutions to direct problems: Given the orbit of the particle and given the center of force relative to that path, one seeks to find the law of force that will maintain the given orbit in terms of the distance *SP* from the particle *P* to the force center *S*.

[Corollary] *Specifically, the solid* $SP^2 \times QT^2 / QR$ *reciprocally proportional to this force must be computed. We shall give examples of this point in the following problems.*

Specifically, from the geometry of the given orbit and force center, one must express the discriminate ratio QR/QT^2 in terms of *SP* and constants of the orbits. The force will then be expressed in terms of the distance *SP* alone.

CONCLUSION

In what follows in the next chapter, Newton gives detailed solutions for three direct problems: Problem 1, find the force that generates a circular orbit with the center of force on the circumference of the circle; Problem 2, find the force that generates an elliptical orbit with the center of force at the center of the ellipse; and finally, Problem 3, find the force that generates a planetary elliptical orbit with the force center (the sun) at a focus of the ellipse. Theorem 3 provides the fundamental paradigm for solving direct problems in which one is given an orbit and a center of force fixed relative to that orbit. The theorem enables one to express the discriminate ratio QR / QT^2 in terms of *SP* and constants of the orbits and thus to determine the nature of the force required to maintain the given orbit about the given center of force. In Theorem 3, Newton fashioned the ratio that measures the force $QR / SP^2 \times QT^2$ from three elements: (1) the parabolic approximation: in the limit as the point Q approaches the point P, then the force is approximately constant in magnitude and direction; (2) Galileo's relationship (in Hypothesis 4): the displacement is proportional to the square of the time for a constant force; and (3) Kepler's relationship (in Theorem 1): the area swept out is proportional to the time for a centripetal force. The result is simplicity itself. From items (1) and (2) the force F is proportional to the deviation QR and inversely proportional to the square of the time t. From item (3) the time t is proportional to the area $SP \times QT$. Thus, the force F is proportional to the linear dynamics ratio $QR / SP^2 \times QT^2$. In the following chapter, this basic result is applied to three specific direct problems, of which the final one is the distinguished Kepler problem of planetary motion.

FIVE

The Paradigm Applied
On Motion, Problems 1, 2, and 3

Theorem 3 in Newton's tract *On Motion* provides the basic paradigm for solutions to direct problems: Given the orbit and the location of the force center, find the force. As Newton put it, "specifically the solid $SP^2 \times QT^2 / QR$ must be computed." He concluded Theorem 3 with the statement, "We shall give examples of this point in the following problems." In this tract, he elected to solve three examples of direct problems. The most important example was Problem 3, the Kepler problem: find the centripetal force required to maintain planetary elliptical motion about a center of force located at the focus of the ellipse. Kepler had demonstrated that a planet P moves in an elliptical orbit about the sun S located at a focal point of the ellipse, and in Theorem 3 Newton demonstrated that the nature of the force required to maintain that motion is inversely proportional to the square of the distance SP, which is the mathematical statement of the law of universal gravitation. Problem 3 clearly has important physical significance. Preliminary to the solution of that very important problem, however, Newton presented the solutions to two other direct problems: find the force that generates (1) circular motion with a center of force on the circumference of the circle and (2) elliptical motion with a center of force at the center of the ellipse (see fig. 5.1). Problems 1 and 2, however, have no clear physical significance, that is, they have no direct application to physical phenomena such as Problem 3 does to the motion of planets.[1] They appear as relatively simple preliminary mathematical exercises in the application of Theorem 3, and they serve only to prepare the reader for the more complex solution that follows them in Problem 3. In an attempt to make clear the general method employed by Newton in applying Theorem 3 to these specific examples, I repeat the suggestion I

96   A GUIDED STUDY

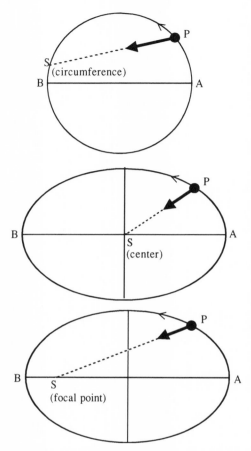

Figure 5.1. In the three direct problems analyzed in *On Motion,* a body *P* moves on a given path *APB* about a given center of force *S*: (1) circular path/circumference, (2) elliptical path/center, and (3) elliptical path/focus.

made in chapter 2 that Newton's exemplar solutions for direct problems be distilled into the following general pattern of analysis:

Step 1. The Diagram. A drawing is provided that identifies the specific orbit corresponding to the general orbit *QPA* in Theorem 3. The immediate position *P* of the body is located, and the line of force *SP* is constructed that connects the body *P* with the force center *S*. Then the future position *Q* of the body is located, and the two lines *QR* and *QT* are constructed. Thus, all three elements of the linear dynamics ratio $QR/QT^2 \times SP^2$ are identified in the diagram.

THE PARADIGM APPLIED

Step 2. The Analysis. Given the full diagram, Newton begins the search for the geometric relationships that will reduce the discriminate ratio QR/QT^2 to a form in which it is expressed as a function of the distance SP alone. It is in this search that Newton displays his command of geometry, conic sections, and mathematical insight; and it is here that the reader must be careful not to lose sight of the general structure of the dynamics in the flurry of mathematical details.

Step 3. The Limit. The general theorem holds only in the limit as the future point Q approaches the immediate point P. Thus, Newton need not search for exact geometric relationships, but only for those that will reduce to the desired functional form in that limit as the point Q approaches the point P. Such relationships will eventually be sought out by others employing the methods of the calculus, but here Newton employed his unique geometric/limiting technique that serves in its stead.

I divide the following detailed discussion of each of the three problems into the three steps just outlined. No such explicit formal division exists in Newton's tract, but I offer it here as a guide for the first-time reader.

PROBLEM 1: A CIRCULAR ORBIT WITH THE CENTER OF FORCE ON THE CIRCUMFERENCE OF THE CIRCLE

As the first example of the application of the paradigm of Theorem 3 to the solution of direct problems, Newton determined the nature of the force required to maintain a circular orbit with the center of force on the circumference of the circle. This problem had no obvious physical application, but it is the simplest of the three examples he presented and thus serves the reader in understanding the more difficult examples to follow.

> Problem 1. *A body orbits on the circumference of a circle; there is required the law of centripetal force being directed to some point on the circumference.*

Step 1: The Diagram

Figure 5.2 is based on the diagram that accompanies this problem in Newton's manuscript. The dynamic elements of Theorem 3 are evident in this figure: the projected tangential displacement PR, the actual circular arc PQ, and the deviation QR. The center of force S is located on the circumference of the circular arc $SQPA$. The chord QL and the normal QT are also elements in the analysis of the force.

> [1-A] *Let* SQPA *be the circumference of a circle,* S *the center of centripetal force,* P *a body carried on the circumference, and* Q *a nearby position into which it will be moved.*

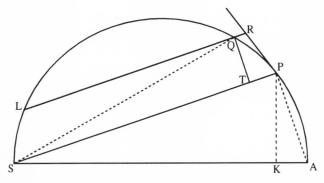

Figure 5.2 Based on Newton's diagram for Problem 1. A body *P* moves in a circular orbit *APQLS* about a center of force *S* located on the circumference of the circle.

The general curve *QPA* from Theorem 3 appears here as a portion of the specific circular path *SQPA* of Problem 1.

[1-B] *To the diameter* SA *and to* SP *drop the perpendiculars* PK *and* QT,

The line *QT*, which is constructed perpendicular to the line *SP*, is one of the two elements in the discriminate ratio QR / QT^2 from Theorem 3, which is needed to express the force law in terms of the radius *SP*.

[1-C] *and through* Q *draw* LR *parallel to* SP, *reaching the circle at* L *and the tangent* PR *at* R.

The second element of the determinate ratio *QR* and the chord of the circle *QL* are defined.

Step 2: The Analysis

In what follows, Newton expressed the discriminate ratio QR / QT^2 in terms of the radius *SP* and the given constant diameter of the circle *SA*.

1. He first demonstrated from similar triangles that $SA^2 / SP^2 = RP^2 / QT^2$.
2. Then he called upon a proposition from Euclid to show that $RP^2 = QR \times LR$.
3. Finally, he argued that the line *LR* can be replaced by the line *SP* in the limit as point *Q* approaches point *P*.

Thus, $SA^2 / SP^2 = RP^2 / QT^2 = (QR \times LR) / QT^2 = (QR \times SP) / QT^2$, which can be solved for the discriminate ratio (i.e., $QR / QT^2 = SA^2 / SP^3$). When that is done, the linear dynamics ratio $QR / (QT^2 \times SP^2)$, and hence the force *F*, can be expressed in terms of *SP*.

THE PARADIGM APPLIED   99

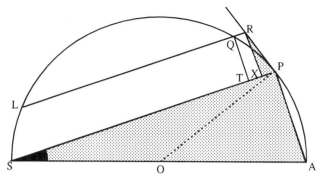

Figure 5.3A A revised diagram for Problem 1. The perpendicular *RX* and the radius *OP* are added.

Figure 5.3B The triangle *RPX* is similar to the triangle *SAP*.

1. From Theorem 3, $F \propto QR / (QT^2 \times SP^2) = (QR / QT^2)(1 / SP^2)$.
2. Substituting for QR / QT^2 from above,
   $F \propto (SA^2 / SP^3)(1 / SP^2) = SA^2 / SP^5$.
3. Because *SA* is the given constant diameter of the circle, the force is inversely proportional to the fifth power of the radius, that is, $F \propto 1 / SP^5$.

Thus, the proportional dependence of the force *F* on *SP* is known and so the solution to Problem 1 is given (i.e., $F \propto 1 / SP^5$). What follows is Newton's detailed analysis for this problem.

[2-D] *There will be* RP^2 *(that is,* $QR \times LR$) *to* QT^2 *as* SA^2 *to* SP^2.

First, one must demonstrate that $RP^2 / QT^2 = SA^2 / SP^2$. See figure 5.3A for a revised diagram[2] that defines another perpendicular to *SP* at a point X. From the similar triangles *RXP* and *SPA* (see above) one has $RP / RX = SA / SP$, and from the parallelogram *RPXQ* one has $RX = QT$. Thus, $RP / QT = SA / SP$. Newton expresses the ratio $RP^2 / QT^2 = SA^2/SP^2$ as squares because QT^2 is needed in the discriminate ratio, QR / QT^2.

Similarity: (1) Angle $OPR = 90°$ because *OP* is a radius of the circle and *PR* is the tangent to the circle. Angle $PXR = 90°$ by construction. Thus,

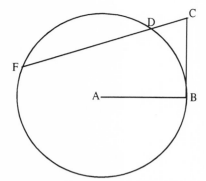
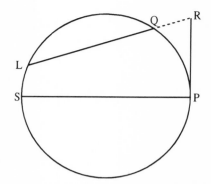

Figure 5.4A From Proposition 36 of Book 3 of Euclid's *Elements*: $CF/CB = CB/CD$.

Figure 5.4B Thus, $RL/RP = RP/QR$ or $RP^2 = QR \times RL$ as required in Problem 1.

angle SPR is the complement of both angles OPS and PRX and thus angles OPS and PRX are equal.

(2) Angle PSO = angle OPS because OS and OP are both radii of the circle. Thus, angle PSO = angle PRX.

(3) Angle $APS = 90°$ because PS and PA are chords of the circle terminating on the diameter SA. Thus, triangles RXP and SPA have all angles equal and are thus similar. Note in figure 5.3B that line RX equals line QT and thus the ratio $RP/RX = RP/QT = SA/SP$ as required above.

Second, one must demonstrate that $RP^2 = QR \times LR$, a relationship that introduces the deviation QR into the analysis. Newton obtained this result from Proposition 36 of Book 3 of Euclid's *Elements* to relate the following lines (see fig. 5.4A): CB (the tangent), CD (the deviation), and CF (the chord plus the deviation). Euclid's basic relationship is given as $CF/CB = CB/CD$, or $BC^2 = CD \times CF$. Figure 5.4B extends Euclid's relationship to the diagram for Problem 1, where $CB = RP$, $CD = QR$, and $CF = LR$. Thus, $QR/RP = RP/LR$, or $RP^2 = QR \times LR$, as required above.

[2-E]   *Therefore* $(QR \times LR) \times (SP^2/SA^2) = QT^2$.

The first result from [2-D] gives $RP^2/QT^2 = SA^2/SP^2$ or $QT^2 = (SP^2/SA^2)RP^2$. The second result from [2-D] gives $RP^2 = (QR \times LR)$. Substituting the second value for RP^2 into the first expression gives the desired result (i.e., $QT^2 = (QR \times LR)(SP^2/SA^2)$). Thus, Newton has obtained an expression relating the two elements QT^2 and QR of the discriminate ratio QR/QT^2, which is required in the linear dynamics ratio $QR/(QT^2 \times SP^2)$ to measure the force.

THE PARADIGM APPLIED

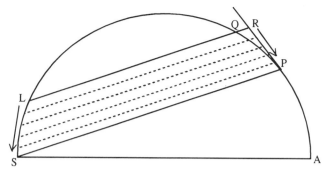

Figure 5.5 As the point Q approaches the point P, the line segment LR approaches the line segment SP.

Step 3: The Limit

[3-F] *Multiply these equals by SP^2 / QR, and, with the points P and Q coalescing, let SP be written in place of LR. Thus, $SP^5 / SA^2 = QT^2 \times SP^2 / QR$.*

From [2-E], $QT^2 = (QR \times LR)(SP^2 / SA^2)$. The linear dynamics ratio can be obtained if both sides of the expression above are multiplied by SP^2 / QR (i.e., $QT^2 \times SP^2 / QR = (QR \times LR)(SP^2 / SA^2)(SP^2 / QR)$) which, upon canceling QR and combining powers of SP, reduces to $QT^2 \times SP^2 / QR = LR \times SP^4 / SA^2$. In the limit, "with the points P and Q coalescing," $Q \to P$, then $LR \to SP$ (see fig. 5.5). When SP is substituted for LR, the expression above reduces to $QT^2 \times SP^2 / QR = LR \times SP^4 / SA^2 \to SP^5 / SA^2$, or $SP^5 / SA^2 = QT^2 \times SP^2 / QR$ as required above.

Conclusion

[3-G] *Therefore the centripetal force is reciprocally as SP^5 / SA^2, that is (because SA^2 is given), as the fifth power of the distance SP. Which was to be proven.*

Or, $F \propto 1/SP^5$. The centripetal force F is given by the linear dynamics ratio $QR / (SP^2 \times QT^2)$, which from [3-F] is equal to (SA^2 / SP^5). Since SA is the given constant diameter of the circle, the force F is inversely proportional to the fifth power of the distance SP and the solution to Problem 1 is given (i.e., $F \propto 1 / SP^5$).

Scholium

In this case and in other similar cases, it must be understood that after the body reaches the center S, it will no longer return to its orbit, but it will depart along the tangent.

Recall that the force center was located on the circumference of the circle and therefore the body will eventually pass through the force center. In such a singular situation both the force and speed would increase without limit and, as Newton states, the body would shoot off along the tangent. The statement is not carried through to the 1687 edition of the *Principia*.

> *In a spiral which cuts all the radii at a given angle, the centripetal force being directed to the beginning of the spiral is reciprocally in the tripled ratio of the distance, but at that beginning no straight line in a fixed position touches the spiral.*

A full solution to the problem of the equiangular spiral appears as Proposition 9 in the 1687 edition of the *Principia* with the force center at the pole (or "beginning") of the spiral. At the pole, no tangent can be defined, but elsewhere the force is proportional to the inverse cube of the distance *SP*.

Newton has thus demonstrated that the centripetal force required to maintain a circular orbit with a center of force located on the circumference of that circle is inversely proportional to the fifth power of the distance *SP*. As has been previously noted, such an orbit with such a center of force does not serve any particular physical situation. The example is intended purely as a demonstration of how the general paradigm in Theorem 3 can be applied to a very simple problem. The analysis requires only one set of similar triangles and a single reference to a proposition from Euclid. (In contrast, the analysis for the Kepler problem in Problem 3 has some sixteen sub-steps.) The particular example of Problem 1 is carried forward to the 1687 edition of the *Principia* in much the same form as given in the tract *On Motion*.

PROBLEM 2: AN ELLIPTICAL ORBIT WITH THE CENTER OF FORCE AT THE CENTER OF THE ELLIPSE

As the second example of the application of the paradigm of Theorem 3 to direct problems, Newton determined the nature of the centripetal force required to maintain an elliptical orbit with the force center located at the *center* of the ellipse. Figure 5.6 is based on the diagram that Newton provided for this problem. As in the previous problem, the dynamic elements of Theorem 3 can be identified: the projected tangential path *PR*, the elliptical arc *PQ*, and the deviation *QR*, here constructed correctly as being parallel to the line of force *PC*. I break Newton's statement of the problem and its solution down into three steps as outlined earlier and provide a line-by-line commentary. I have not given a discussion of Problem 2 as detailed as that in the preceding problem or as detailed as that I shall give in the distinguished Kepler problem (Problem 3). Readers may

THE PARADIGM APPLIED

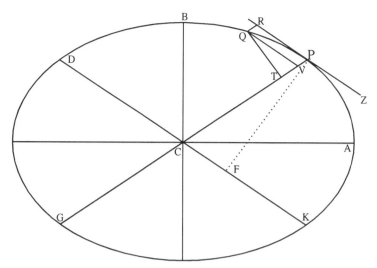

Figure 5.6 Based on Newton's diagram for Problem 2. A body *P* moves on an elliptical orbit *APQB* about a center of force *C* located at the center of the ellipse.

want simply to follow the flow of the solution to Problem 2 and reserve their energy for the solution to Problem 3.

Problem 2. *A body orbits on a classical ellipse; there is required the law of centripetal force being directed to the center of the ellipse.*

Step 1: The Diagram

[Problem 2] *Let* CA *and* CB *be the semi-axes of the ellipse;* GP *and* DK *conjugate diameters;* PF *and* QT *perpendiculars to these diameters;* QV *ordinate to the diameter* GP; *and* QVPR *a parallelogram.*

[1-A] *Let* CA *and* CB *be the semi-axes of the ellipse;*

The semi-major axis *CA* and the semi-minor axis *CB* are constructed perpendicular to each other (see fig. 5.6). Note also that the conjugate diameters *GP* and *DK* (defined next) are not, in general, mutually perpendicular.

[1-B] GP *and* DK *conjugate diameters;*

The transverse diameter *GP* is constructed from the point *P* through the center *C* of the ellipse to the point *G* opposite point *P* (see fig. 5.6). The conjugate diameter *DK* is constructed parallel to the tangent *PR*, and it passes through the center *C*. The diameters *GP* and *DK* are said to be a conjugal (united) pair. Many properties of the ellipse can be expressed in terms of these conjugate diameters.

[1-C] PF *and* QT *perpendiculars to these diameters;*

PF is normal to DK and will be used in the calculation of the circumscribed area to the ellipse given in Lemma 1. The line QT is constructed perpendicular to the diameter GP and will be used in the discriminate ratio QR/QT^2 (see fig. 5.7).

[1-D] QV *ordinate to the diameter* GP;

QV is constructed parallel to the tangent PR and hence is parallel to the conjugate diameter DK and furthermore is said to be ordinate to the transverse diameter GP (see fig. 5.7).

[1-E] *and* QVPR *a parallelogram.*

The deviation QR must be constructed parallel to the line of force PC, and is thus parallel to the conjugate diameter PV. The line segment QV is parallel to the tangent RP from [1-D]. Thus, QVPR is a parallelogram (see fig. 5.7).

Step 2: The Analysis

In the analysis to follow, Newton expresses the discriminate ratio QR/QT^2 in terms of the given constant circumscribed area of the ellipse ($BC \times AC$) and the radius PC. Note that the particular radius PC from Problem 2 is equal to the general radius SP from Theorem 3. When the discriminate ratio was determined in terms of PC, then the linear dynamics ratio was also expressed in terms of PC. Thus, the proportional dependence of the force F on PC was known, and so the direct problem was solved ($F \propto QR/(QT^2 \times PC^2)$).

[2-A] *After these have been constructed, there will be [from the* Conics*]* PV x VG *to* QV^2 *as* PC^2 *to* CD^2 *and* $QV^2 / QT^2 = PC^2 / PF^2$, *and on combining these proportions* PV x VG / $QT^2 = PC^2 / (CD^2 \times PF^2) / PC^2$.

Find QT^2: An expression for the line QT is obtained from the similarity of triangles QTV and PFC (i.e., $QT/QV = PF/PC$ or $QT^2 = QV^2(PF^2/PC^2)$). Then, a relationship from Apollonius (i.e., $PV \times VG/QV^2 = PC^2/CD^2$ or $QV^2 = (PV \times VG)(CD^2/PC^2)$) is used to eliminate QV^2 from the expression for QT^2 (i.e., $QT^2 = (PV \times VG)(PC^2/CD^2)(PC^2/PF^2)$).

[2-B] *Write* QR *in place of* PV *and* BC x CA *in place of* CD x PF,

Find QR: The relationship $PV = QR$ (from the parallelogram QRPV) is used to introduce the deviation QR into the expression for QT^2 and to obtain

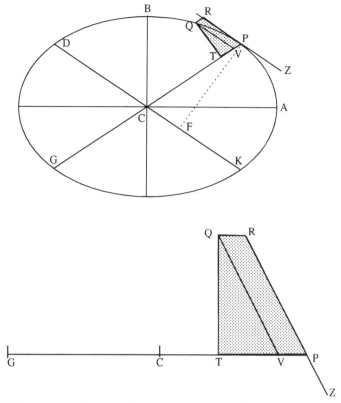

Figure 5.7 The bottom figure is abstracted from the full figure above. The line QT is constructed perpendicular to the conjugate diameter GP, and the line QV is constructed parallel to the tangent ZPR.

the following expression: $QR/QT^2 = PC^4 / (VG)(CD \times PF)^2$, where $(CD \times PF)$ is a constant equal to area $(CA \times CB)$.

Step 3: The Limit

[3-A] *and in addition (with the points P and Q coalescing) 2PC in place of VG, and, when the ends and middles are multiplied into each other, there will result* $QT^2 \times PC^2 / QR = 2BC^2 \times CA^2 / PC$.

Find $QR/(QT^2 \times PC^2)$: Finally, when the point Q approached the point P, then the line VG approached the value $2PC$, and the expression for the discriminate ratio becomes $QR/QT^2 = PC^3 / 2(CA \times CB)^2$.

[3-B] *The centripetal force is therefore reciprocally as* $2BC^2 \times CA^2 / PC$, *that is (because $2BC^2 \times CA^2$ is given), as $1 / PC$, that is, directly as the distance* PC. *Which was to be found.*

Dividing both sides by PC^2 gives the linear dynamics ratio $QR / (QT^2 \times PC^2) = PC / 2(CA \times CB)^2$, which, because the area $(CA \times CB)$ is a constant, gives the ratio as directly proportional to PC.

Conclusion

From Theorem 3, $F \propto QR / (QT^2 \times PC^2) \propto PC$, or the centripetal force F required to maintain elliptical motion about a center of force located at the center C of the ellipse is directly proportional to the distance PC, which is the solution for Problem 2. Newton has thus demonstrated that the centripetal force required to maintain an elliptical orbit about a center of force located at the center of the ellipse is directly proportional to the first power of the distance PC. As noted earlier, such an orbit with such a center of force did not represent any particular physical situation. Newton intended this example purely as a demonstration of how the general paradigm in Theorem 3 was to be applied to yet another direct problem. The analysis for Problem 2 is more complicated than that for Problem 1, but it is relatively simpler than that for Problem 3. The particular example of Problem 2 is carried forward to the 1687 edition of the *Principia* in much the same form as given above for the tract *On Motion*.

PROBLEM 3: AN ELLIPTICAL ORBIT WITH THE CENTER OF FORCE AT A FOCUS OF THE ELLIPSE

The challenge set forth in Problem 3 is a distinguished one: find the solution to the "Problem of the Planets." Kepler had demonstrated in 1609 that Mars moved in an elliptical orbit with the sun at a focus. The question that remained to be answered, however, was the mathematical nature of the force required to maintain that motion. In *On Motion*, the solution follows the two preliminary examples without any fanfare. In the *Principia*, however, Newton called attention to "the dignity of the problem and its use in what follows" by separating Problem 3 from the preceding direct problems and giving it a place of honor at the beginning of a new section. It is the keystone of both works in terms of the dignity of the problem and in the degree of mathematical difficulty.

As in all the previous problems, Newton's diagram identifies the dynamic elements to be employed in the demonstration (see fig. 5.8 in the following statement of the problem): the projected tangential path *PR*, the elliptical arc *PQ*, and the deviation *QR* from the point *R* on the tangential path to the point *Q* on the elliptical arc. Note in this diagram that the deviation QR is constructed parallel to the line of force SP, which is directed toward the focus S, and that the line segment QT is constructed perpendicular to SP. The challenge for this problem is, as it was for the two previous problems, to express the discriminate ratio of the force QR / QT^2 in terms of the radius SP and/or constants of the orbit. The final expression of the solution appears deceptively simple: the discriminate ratio QR / QT^2 is found to be proportional to the reciprocal of the constant *latus rectum* L of the ellipse (where $L = 2BC / AC$), and therefore the force is proportional to the inverse square of the distance, $1 / SP^2$.

$$F \propto QR / (QT^2 \times SP^2) \propto 1 / (L \times SP^2) \propto 1 / SP^2$$

where L is a constant of the ellipse. Thus, once it is demonstrated that the discriminate ratio QR / QT^2 is a constant for the elliptical/focal motion, then the problem is solved. That initial demonstration, however, requires a number of steps, and the reader must remember that the goal is to demonstrate that the discriminate ratio QR / QT^2 is a constant. Once that goal has been achieved, it is a simple matter to determine from Theorem 3 that the force is inversely proportional to the inverse square of the distance SP.

I have divided Step 2, the analysis, into some sixteen sub-steps between line [2-B] and line [2-Q]. The journey is not for the faint of heart; but then if it had not been challenging there would have been no reward for Newton. In honor of the dignity of the problem, I give the statement of the entire proposition and follow it with a line-by-line analysis. I suggest reading the full statement to get an overview of the problem without attempting to justify each point and then following the details of the proof in the line-by-line analysis.

Problem 3. *A body orbits on an ellipse; there is required the law of centripetal force directed to a focus of the ellipse.*

Let S be a focus of the ellipse above. Draw SP cutting the diameter of the ellipse DK at E. It is clear that EP is equal to the semi-major axis AC, seeing that, when from the other focus H of the ellipse the line HI is drawn parallel to CE, because CS and CH are equal, ES and EI are equal, and hence EP becomes half the sum of PS and PI, that is, of PS and PH which are conjointly equal to the total axis 2AC. [See fig. 5.8.]

Let drop the perpendicular QT to SP, and, after calling the principal latus rectum (or $2BC^2 / AC$) of the ellipse L, there will be

L x QR *to* L x PV *as* QR *to* PV, *that is, as* PE (*or* AC) *to* PC;
and L x PV *to* GV x VP *as* L *to* GV;
and GV x VP *to* QV^2 *as* CP^2 *to* CD^2;
and QV^2 *to* QX^2 *as, say,* M *to* N;
and QX^2 *is to* QT^2 *as* EP^2 *to* PF^2, *that is, as* CA^2 *to* PF^2, *or as* CD^2 *to* CB^2.

And *when all these ratios are combined,*

L x QR / QT^2 *will be equal to* (AC / PC) x (L / GV) x (CP^2 / CD^2) x (CD^2 / CB^2),

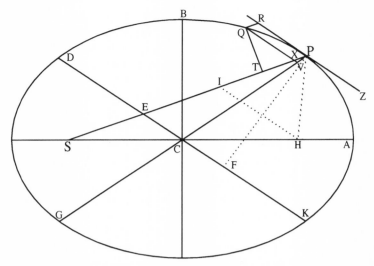

Figure 5.8 Based on Newton's diagram for Problem 3. A body *P* moves in an elliptical orbit *APQB* about a center of force *S* located at a focus of the ellipse.

that is, as AC x L (*or* 2BC²) / (PC x GV) x (CP² / CB²) x (M/N)
or as (2PC / GV) x (M / N).

But with the points P *and* Q *coalescing, the ratios* 2PC / GV *and* M / N *become equal* [*and approach unity*], *and therefore the combined ratio of these* L x QR / QT² [*equals unity (i.e.,* L = QT² / QR)]. *Multiply each part by* SP² / QR *and there will result* L x SP² = SP² x QT² / QR.

Therefore the centripetal force is reciprocally as L x SP², *that is,* [*reciprocally*] *in the doubled ratio of the distance. Which was to be proven.*

The concluding statement of Problem 3 sets forth the mathematical nature of the gravitational force; that is, it is reciprocally as the square of the distance *SP* between the sun and the planet. It is this result that provided the answer to the problem of the planets. I now give a detailed analysis of each portion of Newton's demonstration.

Step 1: The Diagram

[1-A] *Let* S *be a focus of the ellipse above. Draw* SP *cutting the diameter of the ellipse* DK *at* E.

The body is located at a general point *P*, and the center of force is located at a specific point *S*, the focus of the ellipse (see fig. 5.8). The diameter *DK* is one of the conjugate diameters to the point *P*, and it is drawn parallel to the tangent *PR* and through point *C*, the center of the ellipse. The intersection of the diameter *DK* with the line of force *SP* defines the point *E*. Note that the diameter *PG* is not perpendicular to the diameter *DK* but that the line segment *PF* is constructed perpendicular to the diameter *DK*.[3]

Step 2: The Analysis ([2-B] to [2-Q])

In Problem 3, as in the solutions to the two previous direct problems, Newton expressed the discriminate ratio QR / QT^2 in terms of the radius *SP* and/or constants of the figure. When the discriminate ratio was determined in terms of *SP*, then the linear dynamics ratio $QR / QT^2 \times SP^2$ was also expressed in terms of *SP*. So, the proportional dependence of the force *F*, given by the linear dynamics ratio, was also known, and the direct Kepler problem was solved.

As an overview for the reader, I summarize Newton's determination of the discriminate ratio QR / QT^2 here in an effort to provide a general guide to the multiple details in the analysis.

Find *QR*: A relationship for the deviation *QR* is obtained from the similarity of triangles *PXV* and *PEC*, where *QR* = *PX* by construction (i.e., *PE* / *PC* = *PX* / *PV* = *QR* / *PV* or *QR* = *PV*(*PE* / *PC*)). Newton also demonstrated that the line *PE* (defined in fig. 5.8) is equal to the semi-major axis *AC* (a useful relationship for an ellipse that was not found in the standard works on conics). Thus, *QR* can be written as *PV*(*AC* / *PC*). From the same proposition of Apollonius's used in the solution of Problem 2, Newton obtained an expression for *PV* (i.e., $PV \times VG / QV^2 = PC^2 / DC^2$ or $PV = (QV^2 / GV)(PC^2 / DC^2)$), and he used it to eliminate *PV* from the expression for *QR* (i.e., $QR = (QV^2 / GV)(PC^2 / DC^2)(AC / PC)$).

Find QT^2: From the similarity of triangles *EPF* and *XQT*, he obtained an expression for the second element of the discriminate ratio QT^2 (i.e., $QT / QX = PF / PE$ or $QT^2 = QX^2(PF^2 / AC^2)$, where as above *PE* = *AC*). From Euclid's relationship for circumscribed areas (also used in the solution of Problem 2) one has *PF* / *AC* = *BC* / *DC*, and thus $QT^2 = QX^2(BC^2 / DC^2)$.

Find QR / QT^2: The discriminate ratio QR / QT^2 can then be written from $QR = (QV^2 / GV)(PC^2 / DC^2)(AC / PC)$ and $QT^2 = QX^2(BC^2 / DC^2)$ or $QR / QT^2 = (QV^2 / QX^2)(PC / GV)(AC / BC^2)$. Given the definition of the constant *latus rectum* $L = 2BC^2 / AC$, the discriminate ratio becomes $(QV^2 / QX^2)(PC / GV)(2 / L)$.

In the limit as the point *Q* approaches the point *P*, the line *QV* approaches the line *QX*, and the line *GV* approaches a value of 2*PC*. Thus, the discriminate ratio QR / QT^2 approaches $1 / L$, and the linear dynamics ratio $QR /$

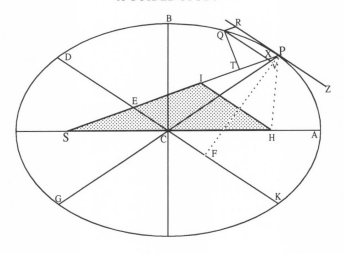

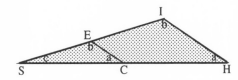

Figure 5.9 The triangle *SIH* below is abstracted from the shaded area of the ellipse above. Line *EC* is parallel to line *IH* and thus triangles *SEC* and *SIH* are similar.

$QT^2 \times SP^2$ approaches $1 / (L \times SP^2)$. The force, therefore, is inversely proportional to the square of the distance *SP*, as was to be demonstrated. What follows is a detailed analysis of Newton's demonstration of the solution.

[2-B] *It is clear that* EP *is equal to the semi-major axis* AC,

At least it will be "clear" that *EP* = *AC* after Newton's demonstration in the next three steps ([2-C] to [2-E]). This particular relationship for an ellipse, *EP* = *AC*, is not found in the standard works on conic sections and appears to be original with Newton.[4]

[2-C] *seeing that, when from the other focus* H *of the ellipse the line* HI *is drawn parallel to* CE, *because* CS *and* CH *are equal,* ES *and* EI *are equal.*

Or, *ES* = *EI*. The lines *CS* and *CH* are equal because they both locate a focus (*S* or *H*) relative to the center *C*. The equality of *ES* and *EI* follows from the similarity of triangles *SEC* and *SIH* and the equality of line segments *CS* and *CH* (see fig. 5.9).

THE PARADIGM APPLIED

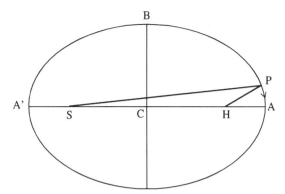

Figure 5.10 The sum $(SP + PH)$ is a constant for any given ellipse, and when the point P goes to A that sum is equal to $2AC$.

Similarity: *HI* was constructed parallel to *CE*, and these two parallel lines are cut by the two transversals *SCH* and *SEI*, as demonstrated in figure 5.9 by the small diagram abstracted from the full diagram. Hence, all the angles a, b, and c in triangles *SEC* and *SIH* are equal and thus the triangles are similar.

Because the triangles are similar, C bisects SH, then E bisects SI, and thus $ES = EI$, as was to be demonstrated.

[2-D] *and hence* EP *becomes half the sum of* PS *and* PI,

Or, $EP = \frac{1}{2}(PS + PI)$. From figure 5.9, $PS = EP + ES = EP + EI$ (because $ES = EI$ from [2-C]). Moreover, $PS + PI = (EP + EI) + PI = EP + (EI + PI) = 2EP$. Thus, $EP = \frac{1}{2}(PS + PI)$.

[2-E] *that is, of* PS *and* PH *which are conjointly equal to the total axis* 2AC.

Or, $2EP = (PS + PI) = (PS + PH) = 2AC$. From the properties of conics (Apollonius, Proposition 48, Book 3)[5] the angles made by the tangent and focal lines (i.e., angles *RPS* and *APH*) are equal, and since the line *IH* is constructed parallel to the tangent *RPZ*, the line $PI = PH$. An ellipse is a curve such that the sum of the distances from two fixed points (the foci) to a general point is given (i.e., the sum $(PS + PH)$ is a constant). When the general point P is on the major axis A (see fig. 5.10) then the sum $(PS + PH)$ equals the sum $(AS + AH)$ equals $(2AC)$ because $AH = A'S$. Thus, $(PS + PH) = 2AC$. Therefore, from [2-D] $2EP = (PS + PI) = (PS + PH) = 2AC$, or "clearly" $EP = AC$, as was to be demonstrated in [2-B].

[2-F] *Let drop the perpendicular* QT *to* SP, *and, after calling the principal latus rectum* (*or* $2BC^2 / AC$) *of the ellipse* L,

In [2-F], Newton calls upon the Apollonian definition of the *latus rectum L*, which states that the *latus rectum L* is in the same ratio to the minor diameter $2BC$ as the minor diameter $2BC$ is to the major diameter $2AC$ (i.e., $L : 2BC :: 2BC : 2AC$ or $L / 2BC = 2BC / 2AC$) or $L = 2BC^2 / AC$ as given above.

[2-G] *there will be* L x QR *to* L x PV *as* QR *to* PV,

The deviation QR is introduced in the format $(L \times QR) / (L \times PV) = (QR) / (PV)$, which will be extended next. Given $(QR) / (PV)$, then simply multiply by (L / L), which can be written as $(L / L)(QR / PV) = (L \times QR) / (L \times PV) = (QR)/(PV)$.

[2-H] *that is, as* PE (*or* AC) *to* PC;

Or, $L(QR) / L(PV) = AC/PC$. The expression for the deviation QR is extended. Triangles *PEC* and *PXV* are similar because *XV* and *EC* are parallel lines (see fig. 5.11). Thus, from the ratio of sides of the similar triangles, $PX / PV = PE / PC$. Moreover, $PX = QR$ because *QRPX* is a parallelogram, and $PE = AC$, as demonstrated in [2-E]. Therefore, the ratio of sides can be written as $QR / PV = AC/PC$. Thus, from this result and [2-H], one obtains $L(QR) / L(PV) = AC / PC$, as required.

[2-I] *and* L x PV *to* GV x VP *as* L *to* GV

The element *PV* is introduced in a similar format to that employed to introduce *QR* in [2-G]. Given L / GV, simply multiply by PV / PV, which can be written as $(PV/ PV)(L / GV) = (L \times PV) / (GV \times VP) = L / GV$.

[2-J] *and* GV x VP *to* QV² *as* CP² *to* CD²;

Or, $GV \times VP / QV^2 = PC^2 / DC^2$. This property of an ellipse will be used to eliminate *PV* from the expression for *QR*. The particular reference is to Proposition 15 of Book 1 of Apollonius's *Conics*, in which it is demonstrated that any chord parallel to a diameter of an ellipse is bisected by the conjugate diameter and, moreover, its square is equal to the product of the portions of the conjugate diameter. In figure 5.12, if the chord is given by *QVQ'* and the conjugate diameter by *PVG*, then $(PV \times VG) = (QV \times VQ')$, where $QV = VQ'$. The same relationship holds for the chord DK, that is $(DC \times CK) = (BC \times CG)$, where $DC = CK$ and $BC = CG$. Thus, the ratio $PV \times VG / QV^2$ about point *V* is equal to the ratio PC^2 / CD^2 about point *C*, as required above.

[2-K] *and* QV² *to* QX² *as, say,* M *to* N;

The ratio M / N is simply a definition of the ratio of QV / QX, a step that Newton eliminates in the 1687 edition of the *Principia*.

[2-L] *and* QX² *is to* QT² *as* EP² *to* PF²,

THE PARADIGM APPLIED

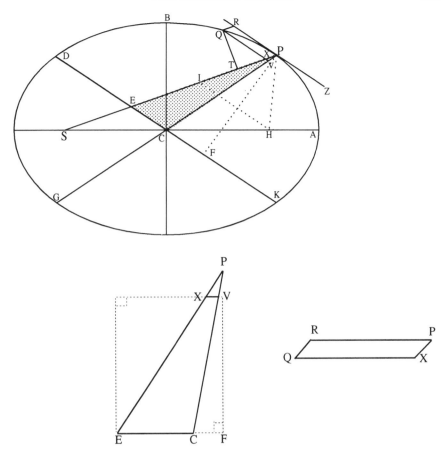

Figure 5.11 The shaded area *CEP* from the ellipse above is abstracted as the similar triangles *PXV* and *BEC* below. The parallelogram *QRPX* from the ellipse is shown enlarged below.

The second element of the discriminate ratio QT^2 is now introduced. Because triangle *EPF* is similar to triangle *XQT* (see fig. 5.13), the ratio of similar sides gives the ratio $QX/QT = EP/PF$.

Similarity: Lines *QX* and *EF* are parallel lines cut by the transversal *EP*, as demonstrated in the small drawing abstracted from the full drawing in figure 5.13. Thus, angles *PEF* and *QXT* are equal, as are the right angles *EFP* and *XQT*. Thus, all the angles of triangles *EPF* and *XQT* are equal and the triangles are similar.

Thus, $QX/QT = EP/PF$ and as given above, $QX^2/QT^2 = EP^2/PF^2$.

114 A GUIDED STUDY

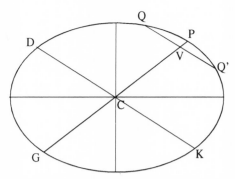

Figure 5.12 The diameter *PG* bisects the chords *QQ'* and *DK*. From Proposition 15 of Book 1 of Apollonius's *Conics*, the ratio of $PV \times VG / QV^2$ is equal to the ratio PC^2 / DC^2.

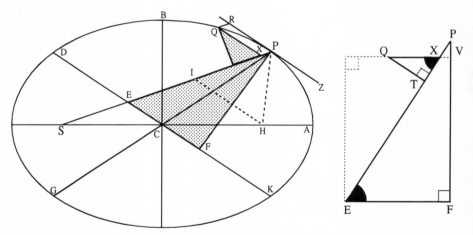

Figure 5.13 The shaded area in the ellipse is abstracted as the triangles *EPF* and *XQT*. Since the lines *QX* and *EF* are parallel, the triangles are similar.

[2-M] *that is, as* CA^2 *to* PF^2, *or as* CD^2 *to* CB^2. [*Finally, that is,* QX^2 *to* QT^2 *as* CD^2 *to* CB^2.]

From [2-B] or [2-E], $EP = CA$, thus $EP / PF = CA / PF$. Also, from Lemma 1, area $4(CA)(CB)$ = area $4(CD)(PF)$ (see fig. 5.14), or $CA / PF = CD / CB$. Thus, from [2-L] $QX / QT = EP / PF = CA / PF = CD / CB$, which can be written in the square as in line [2-L] (i.e., $QX^2 / QT^2 = CD^2 / CB^2$).

THE PARADIGM APPLIED

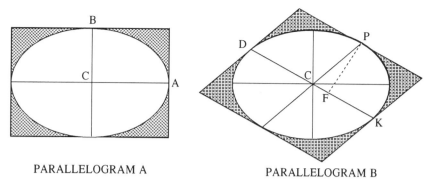

PARALLELOGRAM A   PARALLELOGRAM B

Figure 5.14 The area of parallelogram A (= $4BC \times CA$) is equal to the area of parallelogram B (= $4CD \times PF$).

[2-N] *And when all these ratios are combined,* $L \times QR / QT^2$

Multiply together the five preceding relationships (i.e., lines [2-H] to [2-L]) and consider the products of the left- and right-hand sides separately. The product of the left-hand side of [H][I][J][K][L] is equal to the product $(L \times QR)(L \times PV)(GV \times PV)(QV^2)(QX^2)$ divided by the product $(L \times PV)(GV \times PV)(QV^2)(QX^2)$ (QT^2). By simple cancellation of equals, this ratio of products reduces to $(L \times QR) / QT^2$.

[2-O] *will be equal to* $(AC / PC) \times (L / GV) \times (CP^2 / CD^2) \times (M/N) \times (CD^2 / CB^2)$,

The product of the right-hand side of [H][I][J][K][L] equals the product $(AC)(L)(PC^2)(M)(CD^2)$ divided by the product $(PC)(GV)(CD^2)(N)(CB^2)$, which is equal to $(AC/PC) \times (L/GV) \times (CP^2/CD^2) \times (M/N) \times (CD^2/CB^2)$.

[2-P] *that is, as* $AC \times L$ (*or* $2BC^2$) $/ (PC \times GV) \times (CP^2 / CB^2) \times (M/N)$

That is, by simple cancellation of CD^2 in the ratio given in [2-O], one has $(AC)(L)(PC^2)(M)$ divided by $(PC)(GV)(N)(CB^2)$. Also, from the definition of the *latus rectum* L (see [2-F]), $AC \times L = 2BC^2$. Substituting $2BC^2$ for $AC \times L$, the ratio can be further reduced to $(2BC^2)(CP^2)(M)$ divided by $(PC)(GV)(N)(CB^2)$. This ratio can be rearranged into the ratio given above (i.e., $[(2BC^2) / (PC \times GV)] \times [(CP^2 / CB^2) \times (M/N)]$).

[2-Q] *or as* $(2PC / GV) \times (M/N)$ *or* $[(L \times QR)$ *to* $QT^2]$

Upon canceling BC^2 and PC, [2-P] reduces to $(2PC / GV)(M/N)$, as in [2-Q]. Thus, the left-hand side [2-N] equals the right-hand side [2-Q], or $(L \times QR) / QT^2 = (2PC / GV)(M/N)$.

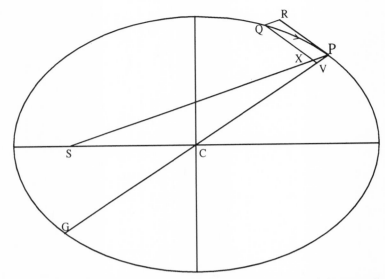

Figure 5.15 As the point Q approaches the point P, the line QV approaches the line QX and the line GV approaches the line GP, which is equal to 2PC.

Step 3: The Limit

[3-R]   *But with the points* P *and* Q *coalescing, the ratios* 2PC / GV *and* M / N *become equal [and approach unity]*,

In the limit as the point Q approaches the point P, then the line GV approaches the line PG = 2PC (see fig. 5.15), and thus the ratio 2PC / GV approaches unity. Also in the limit, the line QV approaches the line QX and from [2-K], $(QV/QX)^2 = M/N$, thus the ratio M / N also approaches unity.

[3-S]   *and therefore the combined ratio of these* L x QR / QT² *[equals unity (i.e.,* L = QT² / QR *)]*

At long last, $QR/QT^2 = 1/L$! Thus, from [2-Q], one has $(L \times QR)/QT^2 = (2PC/GV)(M/N) \to 1$, or $L = QT^2/QR$. Recall that Newton concluded Theorem 3 by stating that the force was to be computed from the ratio $QR/QT^2 \times SP^2$. Everything in this proof has been directed toward showing that the discriminate ratio QR/QT^2 is a constant equal to the reciprocal of the constant *latus rectum L*. The long journey is over.

[3-T]   *Multiply each part by* SP² / QR *and there will result* L x SP² = SP² x QT²/QR.

To obtain the reciprocal of the linear dynamics ratio ($QR / SP^2 \times QT^2$), simply multiply the expression for the discriminate ratio in [3-S] by SP^2. Thus, $SP^2(L) = SP^2(QT^2 / QR) = 1 /$ linear dynamics ratio.

Conclusion

[U] *Therefore the centripetal force is reciprocally as* L $\times$ SP^2, *that is,* [*reciprocally*] *in the doubled ratio of the distance. Which was to be proved.*

From Theorem 3, the force is proportional to the linear dynamics ratio $QR / SP^2 \times QT^2$, or reciprocally as $L(SP^2)$, where the *latus rectum L* is a constant of the ellipse. Thus, the force is proportional to the inverse square of the distance *SP*.

Scholium

[V] *Therefore the major planets orbit in ellipses having a focus at the center of the sun, and with their radii having been constructed to the sun describe areas proportional to the times, exactly as Kepler supposed.*

By 1609, Kepler had obtained both the area law and the general ellipticity of solar planetary orbits by analyzing Tycho's observations of the motion of the planet Mars. In 1664, Newton demonstrated that both relationships can represent motion in the absence of any resistance. Note, however, that Newton demonstrated only that if the planet moves in an ellipse, and if the force center is at a focus, then the force is as the inverse square of the distance (a solution to the direct problem). He has not here demonstrated that if the force is as the inverse square of the distant, then it necessarily follows that the orbit is an ellipse with the force center at a focus of the ellipse (a solution to the inverse problem). In the solution to Problem 4 (to follow) he demonstrates that if the body is initially moving in an arbitrary ellipse under the action of an inverse square force, and if the speed and angle of inclination are changed to take on all possible values, then the resulting motion will be an ellipse, a hyperbola, or a parabola. Newton did not formally address the question of the uniqueness of the solution of the direct problem in the 1678 edition of the *Principia;* however, he did speak to that question in the 1713 edition (to be discussed in chapter 10).

[W] *Moreover, the* latera recta *of these ellipses are* QT^2 / QR, *where the distance between the points* P *and* Q *is the least possible and, as it were, infinitely small.*

For infinitely small displacements the discriminate ratio QT^2 / QR is equal to the constant *latus rectum L*. Newton considers this property of elliptical motion (derived in [3-S]) to be so important that it is elevated to the position of a scholium for easy future reference. Given the detailed analysis required to provide the demonstration, it is a result worthy of the elevation.

CONCLUSION

Thus, Newton demonstrated that the centripetal force required to maintain a body P in elliptical motion about a focal point was inversely proportional to the square of the distance SP. From this solution he developed the concept of a force of universal gravitation that controlled the motion of the planets as they sweep around the sun. All the mathematical astronomers to follow used and extended the solution and, from its beginning here in Newton's reply to Halley's request of 1684, it has developed into the sophisticated gravitational astronomy of the eighteenth, nineteenth, and twentieth centuries. Of all the contributions that Newton made to mathematics, to optics, to astronomy, and to areas other than science, no other received as much public recognition as this solution of the Kepler problem. It elevated both Newton and mathematical astronomy to new heights. One wonders, however, when (if ever) Newton would have made his results public had not Halley happened to visit Cambridge after his discussion with Robert Hooke and Christopher Wren concerning such a proof. As one historian has put it (no pun intended, I am sure), "by 1684, the general question of gravitation was in the air."[6] It is reasonable to assume, therefore, that Newton would eventually have become aware of the general interest in his demonstration. Nevertheless, Halley asked his question at the right time and of the right person, and Newton's answer caught the attention of the academic world.

SIX

The Paradigm Extended

On Motion, Theorem 4 and Problem 4

Contemporary textbooks in astronomy or physics attribute three laws governing planetary motion to the work of Kepler: the first law states that a planet moves in an elliptical path about the sun located at a focus of the ellipse; the second law states that a line joining the sun and a planet sweeps out equal areas in equal times; and the third law states that the period of a planet about the sun is proportional to the three-halves power of the transverse axis of the elliptical orbit. In Theorem 1 of *On Motion,* Newton demonstrates that the second law is valid for any central force; in Problem 3 he demonstrates that, given the first law, it follows that the force will be inversely proportional to the square of the distance; and in Theorem 4, he demonstrates the third law.

THEOREM 4: THE THREE-HALVES POWER LAW

> Theorem 4. *Supposing that the centripetal force is reciprocally proportional to the square of the distance from the center, the squares of the periodic times in ellipses are as the cubes of their transverse axes.*

Figure 6.1 is based on the diagram that accompanies the theorem in the tract *On Motion.* In the following, I consider each line of Newton's demonstration of the theorem in detail.

> [A] *Let* AB *be the transverse axis of an ellipse,* PD *the other axis,* L *the latus rectum,* S *one of the foci;*

Thus, *AB* is the major axis and *PD* is the minor axis. Newton employed the Apollonian definition of the *latus rectum L*, which states that the *latus rectum L* is in the same ratio to the minor diameter *PD* as the minor diameter *PD* is to the major diameter *AB* (i.e., $L : PD :: PD : AB$ or $L / PD = PD / AB$

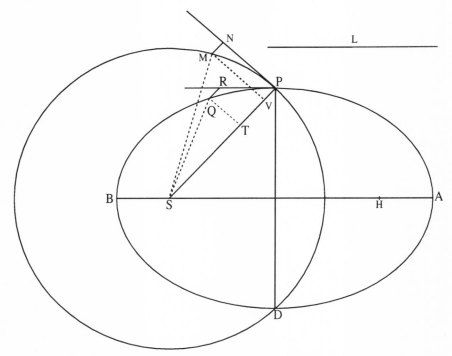

Figure 6.1 Based on Newton's diagram for Theorem 4.

or $L = PD^2 / AB$). From the definition of an ellipse, the distance $(SP + PH)$ is a constant equal to the principal axis AB. In this construction $SP = PH$, therefore $AB = 2SP$ and thus $L = PD^2 / 2SP$, a relationship that will be employed in the following.

[B] *let the circle* PMD *be described with* S *as the center and* SP *as the radius.*

Thus, the circle *PMD* has a radius *SP* and its diameter, $2SP$, is equal to the major axis of the ellipse, *AB*.

[C] *And at the same time let two orbiting bodies describe an elliptical arc* PQ *and the circular arc* PM, *with the centripetal force directed to the focus* S.

It will be demonstrated next that the period of the body orbiting on the circle is equal to the period of the body orbiting on the ellipse when the diameter of the circle is equal to the transverse axis of the ellipse.

[D] *Let* PR *and* PN *be tangent to the ellipse and circle at the point* P. *Parallel to* PS *draw* QR *and* MN *meeting those tangents at* R *and* N.

The deviation *QR* measures the elliptical departure from the tangential inertial motion along *PR* (see fig. 6.2), and the deviation *MN* measures the

THE PARADIGM EXTENDED

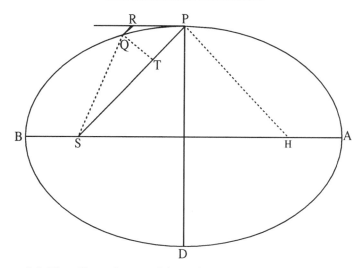

Figure 6.2 The ellipse abstracted from the general diagram for Problem 4.

circular departure from the tangential inertial motion along PN (see fig. 6.3). Both are employed in the linear dynamics ratio as a measure of the force.

[E] *But let the figures PQR and PMN be indefinitely small, so that (by the scholium to Problem 3) there results L × QR = QT² and 2SP × MN = MV².*

In the scholium to Problem 3, Newton demonstrates that the ratio QT^2/QR approaches the *latus rectum* L of the ellipse as the point Q approaches the point P, or ultimately $L \times QR = QT^2$. For a circle, the major axis, the minor axis, and the *latus rectum* are all equal to the diameter, $2SP$. Thus, for the circle, $L = MV^2 / MN = 2SP$ or $2SP \times MN = MV^2$.

[F] *On account of their common distance SP from the center S and therefore equal centripetal forces, MN and QR are equal.*

The centripetal force depends only upon the distance SP, which is the same for the common point P on both orbits (see fig. 6.1). Thus, for a given time, the displacements MN and QR are equal because the forces are equal.

[G] *Consequently QT² is to MV² as L to 2SP, and so QT to MV as the mean proportional between L and 2SP;*

Taking $QT^2 = L \times QR$ and $MV^2 = 2SP \times MN$ from [E], $QT^2/MV^2 = L \times QR / 2SP \times MN = L / 2SP$, since $MN = QR$ from [F]. Thus, $QT^2 / MV^2 = L / 2SP$, or what is equivalent, QT / MV is the mean proportional between L and 2SP.

122   A GUIDED STUDY

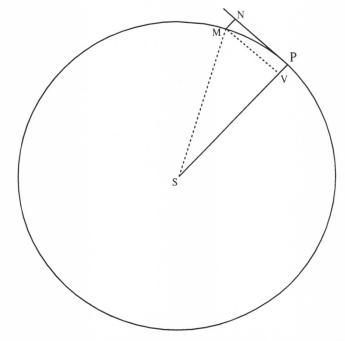

Figure 6.3 The circle abstracted from the general diagram for Problem 4.

[H] [*that is,* PD *to* 2SP]

From [G], $QT^2 / MV^2 = L / (2SP)$. From [A], $L = (PD^2 / AB)$ and $AB = (2SP)$, therefore $L = PD^2 / (2SP)$. Thus, $L / (2SP) = PD^2 / (2SP)^2$. Therefore, $QT / MV = PD / (2SP)$.

  [I] *for this reason the area* SPQ *is to the area* SPM *as the total area of the ellipse to the total area of the circle.*

The area of the ellipse equals $\pi(AB / 2)(PD / 2) = (1/4)\pi AB \times PD$, and the area of the circle equals πSP^2. Hence, the ratio of areas $= AB \times PD / 4SP^2 = PD / 2SP$ because $AB = 2SP$ from line [B]. The ratio of the triangular areas $SPQ = (1/2)SP \times QT$ and $SPM = (1/2)SP \times MV$ reduces to QT / MV, which from line [H], is as $PD / 2SP$. Thus areas $(SPQ / SPM) = PD / 2SP =$ areas (ellipse/circle).

  [J] *But the parts of the areas generated at individual moments are as the areas* SPQ *and* SPM, *and hence as the total areas,*

The incremental areas of the circle ΔA_C and the ellipse ΔA_E are in the same ratio as the total areas given in [I] (i.e., $\Delta A_C / \Delta A_E = A_C / A_E$).

[K] *and consequently when multiplied by the number of moments they will likewise turn out equal to the total areas.*

The total areas equal the sums of the N incremental areas or $A = \Sigma \Delta A = N \Delta A$. Thus, $\Delta A_C / \Delta A_E = (N_C \Delta A_C) / (N_E \Delta A_E) = A_C / A_E$.

[L] *Revolutions on ellipses, therefore, are accomplished at the same time as those on circles whose diameters are equal to the transverse axes of the ellipses.*

From line [K], the number of moments N must be equal (i.e., $N_C = N_E$). Moreover, the number of moments to sweep out the total area equals the period T divided by the size of the equal time increment Δt (i.e., $N = T / \Delta t$). Thus, $N_C / N_E = (T_C / \Delta t) / (T_E / \Delta t) = T_C / T_E$ or $T_C = T_E$.

[M] *The squares of the periodic times in circles (by Corollary 5 of Theorem 2) are as the cubes of their diameters. And hence also in ellipses. Which was to be proven.*

In the 1687 edition, Newton proves this result (Kepler's third law) directly from the limiting relationship given in the scholium to Problem 3 (i.e., $QR / QT^2 \to L$) without using the concentric circle employed here in *On Motion* (1684).

When Newton published the 1687 edition of the *Principia*, which builds directly upon *On Motion*, he adds a third book devoted to the analysis of the actual observations of celestial phenomena. Thus, he brings to life rather abstract mathematical demonstrations of the nature of gravitational force by comparing them to the actual data compiled by astronomers. In the following, Newton adds a scholium to the theorem just discussed, Theorem 4, that suggests one way in which the measurements of the periods of the planets can be used to determine the dimensions of their orbits.

Scholium

Hence in the celestial system from the periodic times of the planets we come to know the proportions of the transverse axes of their orbits. We shall assume one axis, from which others will be given.

The assumed axis usually will be the diameter of the earth's orbit. Then, using Theorem 4, the ratio of the other axes will be known from their observed periods.

[Scholium] *When the axes have been given, however, the orbits will be determined in this way. Let S be the position of the sun, or one focus of the ellipse; A, B, C, D positions of the planets found by observation; and Q the transverse axis of the ellipse. With center A and radius Q – AS let the circle FG be described, and the other focus of the ellipse will be in its circumference.* [See fig. 6.4.]

The points A, B, C, and D all lie on the planetary ellipse. It is a property of an ellipse (see [2-E] in Problem 3) that the sum of the distances from any point on the ellipse to the foci is a constant equal to the length of the

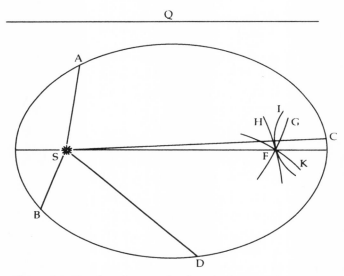

Figure 6.4 The first diagram for the scholium to Theorem 4.

transverse axis, Q. Thus $AS + AF = Q$ or $AF = Q - AS$. Thus, the circle FG of radius $Q - AS$ about the point A will intersect the transverse axis at the focus F.

[Scholium] *Likewise, with centers* B, C, D, *etc., and intervals* Q − BS, Q − CS, Q − DS, *etc., let any number of other circles be described, and that other focus will be in all their circumferences and hence at the common intersection of all of them. If all of the intersections do not coincide, a mean point for the focus must be taken. The advantage of this procedure is that as many observations as possible may be made to elicit a single conclusion, and they may be expeditiously compared with one another.*

The transverse axis Q and the positions A, B, C, D, etc., are only known approximately. Thus, the various circles will not intersect at a single point. If all the points are known with equal accuracy, however, then the mean point for F could be defined as the simple arithmetic mean.[1]

[Scholium] *Halley has shown, however, how to find the individual positions* A, B, C, D, *etc., of a planet from pairs of observations, once the great orbit of the earth is known.*[2] *If that great orbit is not yet considered to be exactly enough determined, then by knowing it approximately, the orbit of any other planet, like Mars, will be determined more closely. Then from the orbit of the earth, the orbit of the planet will be determined more accurately by far than before. And so in turn, until the intersections of the circles in the focus of each orbit concur exactly enough.*

By this method we may determine the orbits of the earth, Mars, Jupiter, and Saturn; but for the orbits of Venus and Mercury, do the following.

THE PARADIGM EXTENDED

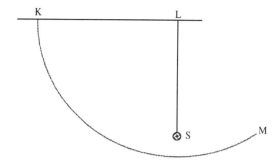

Figure 6.5 The second diagram for the scholium to Theorem 4.

The method Newton describes employs pairs of solar oppositions to fix the position and relative sizes of the planetary radii. Since the two inner planets, Venus and Mercury, can never be in opposition to the sun, then another method, such as the following, must be used for them.

> [Scholium] *From observations made in the greatest digression of the planets from the sun, let tangents of the orbits be obtained. To such a tangent KL let the perpendicular SL be dropped from the sun, and with center L and radius half the axis of the ellipse let the circle KM be described; the center of the ellipse will be in its circumference.* [See fig. 6.5.]

Whiteside suggests that Newton may here assume that his reader would know an equivalent proposition from Apollonius to demonstrate that the center of the ellipse would lie on such a circle.[3]

> [Scholium] *and thus when several circles of this kind are described, it will be found at the intersection of all of them. Then, when the dimensions of the orbits are known, the lengths of these planets will be determined more exactly from their passage through the disc of the sun.*

This technique only gives proportions relative to an assumed size of the earth's orbit. The reference here is to the determination of the absolute size of the distance from the sun to the earth by observations of the transits of Mercury and Venus across the solar disc. Halley had traveled to the island of St. Helena (latitude 16°S) in October of 1677 to observe a transit of Mercury, and he may have called Newton's attention to the possible use of such observations during his famous visit to Cambridge in August of 1684.[4]

PROBLEM 4: GIVEN AN INVERSE SQUARE FORCE, FIND THE RESULTING CONIC SECTION

In Problem 3, the elliptical orbit and focal force center were given, and the force function was found to be inversely proportional to the square of the distance. In Problem 4, the force function is given as inversely proportional to the square of the distance and the orbit is to be determined. Newton considers a body projected at a given point, with a given speed, acted upon by an inverse square force whose absolute value (gravitational constant) is given. From the solution to Problem 3, it is known that one possible orbit under such conditions is that of an ellipse with the center of force at a focus of the ellipse. Newton begins with the assumption that the body is moving in such an ellipse. Moreover, he assumes the existence of an auxiliary circular orbit that is centered on the focus of the ellipse.[5] He then proceeds to determine all the relevant parameters of the initial ellipse from the given elements of the problem. Having done so, he then allows the initial speed of the body to increase and demonstrates that the initial elliptical path becomes a parabolic path and then a hyperbolic path. Newton makes no explicit claim that this solution provides a demonstration that the conic sections exhaust all possible types of motion under the action of an inverse square force (i.e., that it is a solution of the inverse problem). But others saw the possibility of such a solution in his demonstration of Problem 4, which also appears as Proposition 17 in the *Principia*. As Bertoloni Meli points out, support came from both the English mathematician John Keill and the continental mathematician Leonhard Euler.

> In 1716 Keill claimed that Proposition 17 [Problem 4] contained a demonstration of the inverse theorem. In his reply through his student Johann Kruse, however, Johann Bernoulli objected that Proposition 17 [Problem 4] assumes the result rather than proving it. Surprisingly Johann Bernoulli's most talented pupil and possibly the most gifted mathematician of the Enlightenment sided with Keill: Leonhard Euler, in his 1736 *Mechanica,* claimed that the inverse problem of central forces could be solved on the basis of Proposition 17 [Problem 4].[6]

The composite nature of the solution, however, is such that it has been the subject of considerable controversy, and the debate still continues.[7]

I give the statement of the problem and then follow it with a line-by-line analysis of Newton's demonstration of the solution. Figure 6.6 is based on Newton's diagram for Problem 4.

> **Problem 4.** *Supposing that the centripetal force be made reciprocally proportional to the square of the distance from its center, and that the absolute quantity of that force is known; there is required an ellipse which a body will describe, when released from a given position with a given speed along a given straight line.*
>
> [A] *Let the centripetal force directed to point S be that which makes the body* p *orbit in a circle* pq *described with center S and any radius Sp.*

The body p moves in a reference circle pq of radius Sp about the center of force S. For a given "absolute quantity of force" (i.e., a given force con-

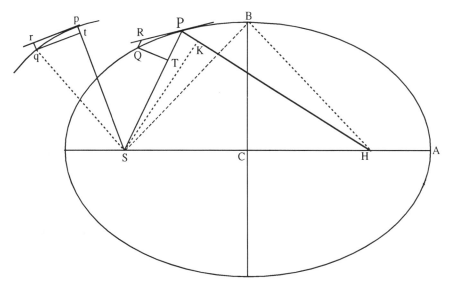

Figure 6.6 Based on Newton's diagram for Problem 4.

stant) there exists a particular constant tangential speed for which the body *p* will move uniformly in a circle, as has been demonstrated in Theorem 2. In what follows, Newton assumes that the proper set of initial conditions has been met and he will relate various aspects of this constant uniform circular motion to the more general conic motion of the orbit PQ (see fig. 6.6).

> [B] *Let the body P be released from the position P along line* PR, *and soon after let it deflect under the compulsion of the centripetal force into the ellipse* PQ. *The straight line* PR, *therefore, will touch this at* P.

This description of the elements of the diagram constitute Newton's standard method of analysis of direct problems. The line *PR* represents the tangential linear displacement the body would have made had no force acted on it. The possibility of elliptical motion under an inverse square centripetal force has been demonstrated in Problem 3, and here Newton assumed that the proper set of initial conditions had been met for the body *P* to move on an ellipse. Moreover, from Theorem 1, the motion will be such that the radius will sweep out equal areas in equal times because the force is directed to a fixed center. Newton was aware that other initial conditions will produce other motions than elliptical ones, and he addressed those alternate motions in what follows.

> [C] *In the same way let the straight line* pr *touch the circle at* p,

The tangent is defined for the reference circle.

[D] *and let* PR *be to* pr *as the initial speed of the body* P *sent out to the uniform speed of the body* p.

The implicit assumption is made that the tangential displacements PR and pr take place in equal times. If v_P is the projection speed at P and v_p is the uniform circular speed at p, then the displacements PR and pr are equal to the product of the speed and the time, or are proportional to those speeds in the same time t (i.e., $PR/pr = v_P t / v_p t = v_P / v_p$).

[E] *Let* RQ *and* rq *be drawn parallel to* SP *and* Sp, *the latter meeting the circle at* q, *the former the ellipse at* Q,

The deviations QR and qr are the first elements required for the respective linear dynamics ratios.

[F] *and let the perpendiculars* QT *and* qt *be drawn from* Q *and* q *to* SP *and* Sp.

The perpendiculars QT and qt are the second elements required for the respective linear dynamics ratios.

[G] RQ *is to* rq *as the centripetal force at* P *is to the centripetal force at* p:

For a given time, the force F_P is proportional to the deviation QR and the force F_p is proportional to the deviation qr, as has been discussed in detail in the previous problem solutions.

[H] *that is, as* Sp² *to* SP², *and hence that ratio is given.*

From the statement of the problem, $F_P / F_p = (C / SP^2) / (C / Sp^2) = Sp^2 / SP^2$, where C is a constant set by the given absolute value of the force. Since the ratio of the displacements RQ / rq is given in [D] by the initial projection speed, and since that ratio equals the ratio of the forces [from line [G]], then ratio Sp^2 / SP^2 is also given, and it equals QR / qr.

[I] *The ratio* QT *to* qt *is also given.*

The ratio of the areas, $(QT \times SP) / (qt \times Sp)$, is equal to the ratio of the times (Theorem 1: Kepler's law of equal areas in equal times). Since the times are equal, then $(QT \times SP) = (qt \times Sp)$, or $QT / qt = Sp / SP$, and the ratio QT / qt is thus given (because the ratio Sp / SP is given in line [H]).

[J] *From this latter ratio doubled let the given ratio* QR *to* qr *be taken away, and there will remain the given ratio of* QT² / QR *to* qt² / qr,

Divide the square of the given ratio of $(QT / qt)^2$ by the given ratio of (QR / qr), and there results the given ratio of $(QT^2 / QR) / (qt^2 / qr)$.

[K] *that is* (*by the scholium to Problem 3*) *the ratio of the* latus rectum *of the ellipse to the diameter of the circle;*

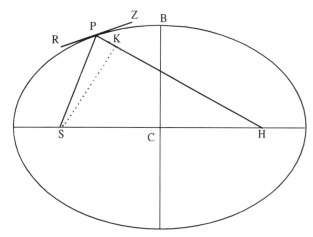

Figure 6.7 Angle *RPS* equals angle *ZPH,* and thus angle *RPH* is the supplement of angle *RPS.*

That is, $(QT^2 / QR) / (qt^2 / qr) = L / Sp$. The scholium to Problem 3 states that the limiting value for the ratio QT^2 / QR for an ellipse as $Q \to P$ is the *latus rectum,* L. Thus, since the ratio in line [J] is known, the ratio of L_E for the ellipse to L_C for the circle is known. The *latus rectum* L_C for the circle, however, is its diameter *Sp.* (From the definition, L = (minor diameter)^2 / (major diameter); for a circle the two diameters are equal; thus, L_C = diameter.)

[L] *and therefore the* latus rectum *of the ellipse is given. Let that be* L.

Thus, since the diameter of the circle *Sp* is given, the *latus rectum L* of the ellipse is known (i.e., from [K] $L / Sp = (QT^2 / QR) / (qt^2 / qr)$, where the final ratio is given from line [J]).

[M] *In addition, the focus of the ellipse* S *is given.*

The focal point *S* is given in the opening statement in line [A].

[N] *Let angle* RPH *be the complement of angle* RPS *to two right angles* [*i.e., angle* RPH *is the supplement of angle* RPS],

From figure 6.7, angle *RPH* + angle *HPZ* = 180°. From the properties of conics (Apollonius, Proposition 48, Book 3) the angles *RPS* and *ZPH* that are made by the tangent and focal lines are equal.[8] Thus, the angle *RPH* = 180° − *ZPH* = 180° − *RPS*, or angle *RPH* is the supplement of angle *RPS.*

[O] *and there will be given in position the line* PH *in which the other focus* H *is located.*

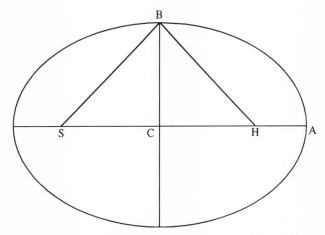

Figure 6.8 The focal distances *SC* and *CH* are equal, and thus *SH* can be given as 2*CH*.

The angle *RPS* is known from the projection velocity. From line [P] the angle *SPH* is known; it is 180° − 2*RPS*. Thus, the direction of the line on which the second focus *H* is located is known relative to the line *SP*, or as Newton puts it, "There will be given in position the line *PH*." The distance *PH*, however, is not yet known. Lines [P] to [V] are required to determine the length of *PH* and thus to locate the second focus *H*.

[P] *After the perpendicular* SK *is let down to* PH *and the semi-minor axis* BC *is erected, there is* SP² − 2KP × PH + PH² = SH²

This result is found in Euclid, Book 2, Proposition 13.[9] (Otherwise, from the law of cosines, *SH*² = *SP*² − 2*SP* × *PH* × cos(angle *SPH*) + *PH*². From figure 6.7, cos(angle *SPH*) = *KP*/ *SP*. Thus, *SP*² − 2*KP* × *PH* + *PH*² = *SH*².)

[Q] SH² = 4BH² − 4BC²

From figure 6.8, *SH* = *SC* + *CH* = 2*CH* or *SH*² = 4*CH*². From the right triangle *BCH*, one has *CH*² = *BH*² − *BC*². Thus, 4*CH*² = 4(*BH*² − *BC*²) or *SH*² = 4*BH*² − 4*BC*².

[R] 4BH² − 4BC² = (SP + PH)² − L × (SP + PH)

From the definition of an ellipse, (*SP* + *PH*) = 2*AC* = 2*BH*, thus 4*BH*² = (*SP* + *PH*)². From the definition of the *latus rectum*, *L* = 2*BC*² / *AC*, thus 4*BC*² = *L* × 2*AC* = *L* × (*SP* + *PH*)². Thus, 4*BH*² − 4*BC*² = (*SP* + *PH*)² − *L* × (*SP* + *PH*) as given above.

[S] (SP + PH)² − [L × (SP + PH)] =
     (SP² + 2SP × PH + PH²) − [L × (SP + PH)].

Simply expand the square, $(SP + PH)^2 = (SP^2 + 2SP \times PH + PH^2)$.

[T] *To each side add* $2KP \times PH + L \times (SP + PH) - SP^2 - PH^2$ *and there will result* $L \times (SP + PH) = 2SP \times PH + 2KP \times PH,$

Starting at line [P], $SP^2 - 2KP \times PH + PH^2$ has been shown to be equal at line [S] to $(SP^2 + 2SP \times PH + PH^2) - L \times (SP + PH)$, which upon cancellation and rearrangement becomes $L \times (SP + PH) = 2SP \times PH + 2KP \times PH$.

[U] *or* $(SP + PH)$ *to* PH *as* $(2SP + 2KP)$ *to* L.

Dividing both sides by $L \times PH$, one has $(SP + PH)/PH = (2SP + 2KP)/L$

[V] *From which the other focus* H *is given.*

The direction of the line PH was determined in line [P]. The length of PH was determined by the ratio in line [U], because the elements, SP, L, and KP are given. ($KP = SP \cos(SPH)$, where the angle SPH is given in line [P].) Thus, the direction and distance of PH relative to S is determined, and therefore the second focus H is known.

[W] *Given the foci, however, along with the transverse axis* $SP + PH$, *the ellipse is given. As was to be proven.*

Knowledge of the position of both foci and the distance $(SP + PH)$ permits construction of the ellipse. A simple demonstration is accomplished by putting a tack at each of the foci and joining them with a loose string of length $(SP + PH)$. When a pencil is moved so that the string remains taut, the ellipse is drawn.

[X] *This argument holds when the figure is an ellipse. But it can happen that a body moves in a parabola or a hyperbola.*

The initial conditions set in line [B] restricted the motion to an ellipse, but in [X] Newton demonstrates that the motion under an inverse square force could also be a parabola or a hyperbola (i.e., any conic section). In doing so, he permits the projection speed v_P of the body P to take on greater values than the specific value that produced elliptical motion at a given point under an inverse square centripetal force with a given force constant. He demonstrates that the resulting motion will be one of the conic sections.

[Y] *But if the speed of a body is so great that the* latus rectum L *is equal to* $2SP + 2KP$, [*then*] *the figure will be a parabola having its focus at the point* S *and all its diameters parallel to the line* PH.

As the second focus H recedes to infinity, the figure becomes a parabola. From line [N], as PH increases without limit, $(SP + PH) / PH \to 1$ or the *latus rectum* L approaches $(2SP + 2PK)$.

[Z] *But if the body is released at an even greater speed, [then] it will be moved in a hyperbola having one focus at the point S, the other at the point H taken on the opposite side of the point P, and its transverse axis equal to the difference of the lines PS and PH.*

The second focus *H*, having receded to infinity, now approaches the other focus *S* from the opposite side of the general point *P*.

Newton did not address the full question of the uniqueness of this solution to Problem 4 (i.e., do the three conic sections exhaust all the possible solutions for motion under an inverse square force?). Perhaps he thought the affirmative answer self-evident, since for every initial speed, angle of projection, and value of force constant, a unique conic is defined: an ellipse, a parabola, or a hyperbola.[10] Moreover, the subject of the uniqueness of the solution was not discussed in the revised solution to Problem 4 that appeared as Proposition 17 in the 1687 edition of the *Principia*. In Propositions 11, 12, and 13, however, he presented the solutions for the direct problems of each of the conic orbits: an ellipse, a hyperbola, and a parabola. In the 1687 edition, he included the following corollary to Proposition 13, which directly assumed uniqueness.

Corollary. *From the last three propositions it follows that if any body P should depart from Position P along any straight line Pr, with any velocity, and is at the same time acted upon by a centripetal force that is reciprocally proportional to the square of the distance from the center, this body will be moved in one of the sections of conics having a focus at the center of forces; and conversely.*

In the 1713 edition, however, he added a statement to the corollary in Proposition 13 that was intended to address and to justify the assumption of uniqueness. (See chapter 10 for a discussion of that revised corollary.) That demonstration is important because, if established, the solution to the direct problem in combination with this solution serves as a solution to the inverse problem. Following Newton, it has been demonstrated to the satisfaction of the most demanding of mathematicians that the conic sections are the unique set of solutions for motion under an inverse square force. In *On Motion* and in the 1687 edition of the *Principia*, however, Newton is satisfied with a much more intuitive demonstration of uniqueness.

In the scholium to Theorem 4, Newton discussed the theorem's application to the computation of the relative size of planetary orbits from the observation of planetary periods. In the scholium to Problem 4, he applies his techniques developed in Problem 4 to the determination of orbits of comets. Again, he applies the ideal mathematical world of the tract to the actual observational world of the heavens. The topic of comets caused Newton much concern. In a letter to Halley, written more than a year after sending the following scholium on comets, Newton states that "in Autumn

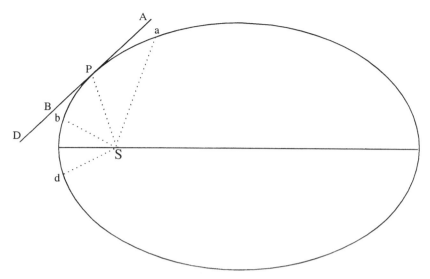

Figure 6.9 The diagram for the scholium to Problem 4.

last I spent two months in calculations to no purpose for want of a good method." Not until Newton replaces the linear approximation discussed next with a parabolic approximation for the elliptical path does the "good method" appear as Proposition 41 of Book Three of the 1687 *Principia*.

Scholium

Now indeed with the help of this problem when it has been solved, it is possible to define the orbits of comets, and from that the times of their revolutions, and from a comparison of the magnitude of their orbits, eccentricities, [perihelia], inclinations to the ecliptic plane and their nodes, to know whether the same comet returns to us frequently.

The most celebrated example of a returning comet is the comet of 1682, which Halley correctly identified as the return of the comet of 1531 and 1607 (a period of 75 or 76 years) and which he correctly predicted to return in 1758 (after his death).

[Scholium] *To be sure, from the four observations of the comet's position, according to the hypothesis that a comet moves in a straight line, its rectilinear path must be determined. Let that [path] be APBD, and let A, P, B, D be positions of the comet on that path at the times of observations, and let S be the position of the sun. Imagine that with the speed at which it regularly traverses the straight line AD, the comet is released from one of its positions P, and, soon forced by centripetal force, it is deflected from the straight path and goes off into the ellipse* Pbda. [See fig. 6.9.][11]

By the time of this analysis of 1684, Newton had come to accept the universality of the gravitational force (i.e., that the force derived in Problem 3 for planetary motion also applied to all celestial bodies, such as comets). Previously there had been a question in Newton's mind concerning the curved nature of cometary paths. The rectilinear hypothesis (that is, comets move in straight lines), however, is simply an approximation: one that will be replaced by a more sophisticated method in the *Principia*.[12]

> [Scholium] *This ellipse must be determined as in the problem above. In it let* a, P, b, d *be positions of the comet at the times of observations. Let the longitudes and latitudes of these positions from the earth be known. As much as the observed longitudes and latitudes are greater or less than these, let new longitudes and latitudes be taken greater or less than the observed ones. From these new [measurements] let the rectilinear path of the comet again be found, and from that the elliptical path as before. And the four new positions on the elliptical path, having been augmented or diminished by previous errors now will agree sufficiently precisely with the observations.*

In the determination of the elliptical orbit in Problem 4 above, one must assume a value for the magnitude of the gravitational force (see the statement of the problem). Moreover, it is not clear how one is to know that the new position will produce convergence with the actual path of the comet. Clearly, Newton is struggling with the procedure.

> [Scholium] *But if perchance palpable errors should still remain, it is possible for the entire task to be repeated. And, so that the computations may not be annoying to the astronomers, it will suffice to determine all these things by a geometrical procedure.*

There then follows another paragraph in which Newton discusses the difficulty in assigning areas proportional to the times. Whiteside observes that the method outlined in this scholium is a "makeshift construction ... more optimistic of a chance success than solidly reasoned."[13] In fact, the entire subject of the motion of comets was a challenge. Newton claimed that "this discussion about comets is the most difficult in the whole book."[14] This challenge continued to receive his attention and would eventually be met in the *Principia*.

In Book Three of the *Principia* Newton will consider many more such observational problems. The following is an example taken from the 1687 *Principia* in which Newton discusses the application of the theorems from Book One to the phenomena discussed in Book Three. In *On Motion,* and in its extension into Book One of the *Principia*, Newton has discussed the ideal case of a single planet orbiting about a single stationary center of force in the sun. In the following proposition from Book Three he discusses the influence of adjacent planets upon their mutual motion (that is, a three-body versus a two-body problem). Now Newton is less a mathematician and more of a practical astronomer. He must call into action estimates of the relative masses of the sun and planets as well as their relative distances. He calls on demonstrations in Propositions 66 and 67 in Book One concerning motion about the center of mass to relate the ideal mathematical problem with a fixed center of force to the actual problem of the sun and planet moving about their center of mass.

BOOK THREE. PROPOSITION 13. THEOREM 13

Book Three. Proposition 13. *Planets move in ellipses having a focus in the center of the sun, and with radii having been constructed to that center they describe areas proportional to the times.*

We have talked above about these motions from the phenomena. Now that we have understood the principles of motions, from these [principles] we infer celestial motions a priori. Since the weights of the Planets toward the sun are reciprocally as the squares of the distances from the center of the sun; if the sun were at rest and the other Planets were not acting on each other, [then] their orbits would be elliptical having the sun in a common focus and areas would be described proportional to the times (by Propositions 1 and 11 and Corollary 1 Proposition 13 Book One [Theorem 1 and Problem 3 in *On Motion*]). *But the actions of the Planets reciprocally on each other are very small (so that they might be able to be neglected), and they disturb the motions of the Planets in ellipses around the sun in motion less (by Proposition 66, Book One) than if those motions were being performed around the sun at rest.*

Indeed, the action of Jupiter upon Saturn must not be altogether neglected. For the gravity toward Jupiter is to the gravity toward the sun (the distances being equal) as 1 to 1,100;

Modern values of the masses of the sun and Jupiter give a ratio of 1 to 1,047, which compares to Newton's value of 1 to 1,100 (a value that appears as 1,067 in the revised *Principia*), values which Newton obtained using Kepler's third law (Theorem 4) and measurements of the periods of their satellites (Venus for the sun, its moons for Jupiter).[15]

[Proposition 13] *and therefore in the conjunction of Jupiter and Saturn, since the distance of Saturn from Jupiter is to the distance of Saturn from the sun, almost as 4 to 9;*

Modern value of the semi-major axis of Jupiter is 5.203 AU, for Saturn it is 9.539 AU, and thus the distance between Jupiter and Saturn at conjunction is the difference, or 4.336 AU. Thus, the ratio of distances is 4.336 to 9.539 or, as Newton states, 4 to 9.

[Proposition 13] *the gravity of Saturn toward Jupiter will be to the gravity of Saturn toward the sun as 81 to 16 × 1,100; or 1 to about 217.*

The gravitational force is directly proportional to the mass and inversely proportional to the square of the distance. Thus, the force on Saturn due to Jupiter is as $1/4^2$ and that due to the sun as $1{,}100/9^2$, or a ratio of $(1/16) / (1{,}100/81) = 81/(16 \times 1{,}100) = 1/217$.

[Proposition 13] *Yet the whole error in the motion of Saturn around the sun, arising from so great a gravity toward Jupiter, can be avoided by constructing a focus of the orbit of Saturn in a common center of gravity of Jupiter and the sun (by Proposition 67, Book One) and hence when the error is greatest it hardly exceeds two minutes. But in the conjunction of Jupiter and Saturn, the accelerative gravities of the sun toward Saturn, of Jupiter toward Saturn, and of Jupiter toward the sun, are almost as 16, 81, and 16 x 81 x 2,360 / 25 or 122,342,*

Using modern notation, the forces between two masses can be expressed as GM_1M_2 / R_{12}^2, where G is the universal gravitational constant and R_{12} is the distance between their centers. The accelerative gravities given are those forces divided by the mass of the force center (that is, the "accelerative gravity" of the sun toward Saturn is $GM_{Saturn}/9^2$, of Jupiter toward Saturn is $GM_{Saturn}/4^2$, and the sun toward Jupiter is $GM_{Jupiter}/5^2$). Thus, the ratios are as $GM_{Saturn}/9^2 : GM_{Saturn}/4^2 : GM_{Jupiter}/5^2$. Newton simplifies the results by dividing by $(GM_{Saturn}) / (9^2 \times 4^2)$, which gives the result $4^2 : 9^2 : (M_{Jupiter} / M_{Saturn}) / 5^2$, where the ratio of $M_{Jupiter} / M_{Saturn}$ is given by 2,360 in the first edition and by 3,021 in the revised editions.

[Proposition 13] *and thus the differences of gravities of the sun toward Saturn, and of Jupiter toward Saturn, is to the gravity of Jupiter toward the sun as 65 to 122,342 or as 1 to 1,867. But the greatest power of Saturn to disturb the motion of Jupiter is proportional to this difference: and hence the perturbation of the orbit of Jupiter is far less than that of Saturn's. The perturbations of the remaining orbits are even far less.*

Thus, does Newton the mathematician and Newton the practical astronomer relate the demonstrations of the theorems and solutions of abstract problems to the observations of the motion of celestial bodies. The applications and examples are increased in the editions of the *Principia* that follow the tract *On Motion*. The demonstration of a universal gravitational force that depends upon the inverse square of the distance is to be found in the heavens as well as in the pages of the *Principia*.

CONCLUSION

Following his receipt in London of Newton's tract *On Motion* in November of 1664, Halley returned to Cambridge to consult with Newton once again. He found Newton occupied with the task of rewriting the work in more detail. That project would eventually require two years to complete. Under Halley's personal encouragement, careful editorial eye, and financial support, it would ultimately result in publication as the first edition of the *Principia* in 1687. Upon his return to London, Halley reported to the Royal Society at its meeting on 10 December 1684, about his interchange with Newton and of his intent to have the tract entered in the *Register* of the

Society, which it was. It was not until two years later, however, at the meeting of the Royal Society on 21 April 1686, that Halley announced that the work was almost ready for the press. At the following meeting, on 28 April 1686, he presented the manuscript of Book One to the society. He delivered Book One to the printer that same month, but it was not until March of the following year that Halley delivered Books Two and Three to the printer.[16] On 5 July 1687, Halley wrote to Newton about the details of the final publication figures and to announce, "I have at length brought your Book to an end."[17]

PART THREE

The Revisions and Extensions to Newton's Solution

Bust of Isaac Newton, by D. Le Marchand, currently located in the library of the Royal Society in London. It is a copy of the original of 1718, which is in the British Museum. Copyright © The Royal Society. Reproduced by permission.

SEVEN

The *Principia* and Its Relationship to *On Motion*
A Reference Guide for the Reader

Just as the seven problems and four theorems of the tract *On Motion* far exceeded Halley's request in 1684 for a solution to the single problem of elliptical motion about a focal center of force, so the published text of the first edition of the *Principia* in 1687 far exceeded the contents of the original tract of 1684. Newton divided the work into three books: the first book is devoted to the analysis of motion in a nonresistive medium, the second book to the analysis of motion in resistive media, and the third book to the analysis of the data for celestial phenomena.

Book One of the *Principia* is concerned with motion in a medium devoid of resistance, as are the first four theorems and problems of the original 1684 tract *On Motion*. These theorems and problems from *On Motion* are contained in Sections 2 and 3 of Book One of the *Principia*. Section 2 starts with the demonstration of Theorem 1 from *On Motion* (Kepler's area law) and Section 3 finishes with the solution of Problem 4 from *On Motion* (the scale of the conic motion). Newton has inserted, however, four additional theorems and five additional problems between these original opening and closing points: what was Theorem 1 in *On Motion* still appears as Proposition 1 in the *Principia*; but what was Problem 4 in *On Motion* now appears as Proposition 17 in the *Principia*. This enlarged version of the initial tract in Sections 2 and 3 is preceded by an extended set of definitions, laws of motion, and lemmas. Moreover, following Sections 2 and 3, Book One contains an additional thirty-one propositions divided into eleven sections that range in subject from finding the conic orbits from a given focus and finding the motions in a given orbit to an analysis of the motion of various pendula.

Book Two of the *Principia* is concerned with motion in a resistive medium. Whereas Problems 1 to 4 of the original 1684 tract *On Motion* are concerned with celestial planetary paths, Problems 5 to 7 are concerned with terrestrial projectile paths. Projectile motion raises the subject of motion in a resistive medium. This subject, which was treated in the two final problems in the tract *On Motion*, grew in Book Two of the *Principia* into nine sections containing a total of fifty-three propositions. Book Two begins in Section 1 with the analysis of the motion of bodies for which the resistance is proportional to the velocity, and it ends in Section 9 with the analysis of the motion of bodies carried in a fluid vortex.

Book Three of the *Principia* treats a subject in detail that in *On Motion* is taken up only briefly: the analysis of celestial and terrestrial data. It begins with the analysis of the data concerning the motion of the moons of the planet Jupiter, discusses the complex data of the motion of the moon of the earth, and concludes with an extended discussion of the data concerning the motion of comets. In the introduction to Book Three, however, Newton advised the reader to begin with only selected sections of Book One and not to attempt to read all the material that precedes Book Three.

> I do not want to suggest that anyone should read all of these Propositions—which appear there in great number—since they could present too great an obstacle even for readers skilled in mathematics. It would be sufficient for someone to read carefully the definitions, laws of motion, and the first three sections of the first book; then let [the reader] skip to this [third] book.[1]

My discussion of the 1684 tract *On Motion* in chapters 4, 5, and 6 contains a detailed commentary on almost every line of text. Taken in conjunction with the detailed commentary on the two pre-1669 tracts in chapter 3, these four chapters here provide a means for reading the first three sections of Book One of the *Principia*. An English translation from the Latin of the first three sections of the 1687 edition of the *Principia* without additional commentary is given in the Appendix. This chapter provides a link between the text of the tract *On Motion* and the text of that translation from the first edition of the *Principia*.

INTRODUCTION

The title page of the *Principia* identifies Newton as the Lucasian Professor of Mathematics at Trinity College, Cambridge, and Samuel Pepys as President of the Royal Society. It is followed by a "Preface to the Reader" that opens with a reference to the ancients and closes with an appeal to the reader to look with patience upon Newton's "labors in a field so difficult." Next there is a dedicatory poem in honor of Newton by the editor, Edmund Halley, which may reveal more about Halley than about Newton. In it Halley compares the gift that Newton has given the world with the gifts of other great contributors, one being the gift of wine from one "who pressed from grapes the mitigation of cares" (a reflection on the values of the socially active Halley rather than those of the reclusive Newton).

DEFINITIONS

The body of the text begins with a series of eight definitions and a long scholium that is an essay on absolute time, space, and motion. Only Definitions 3 and 5 have a counterpart in the original tract *On Motion*.

> *Principia*: Definition 1. *The quantity of matter is the measure of it arising from the density and volume conjointly.*

> *Principia*: Definition 2. *The quantity of motion is the measure of it arising from the velocity and quantity of matter conjointly.*

> *Principia*: Definition 3.
> *The force innate in matter is the power of resisting, whereby each individual body, as much as there is in it to do so, perseveres in its state either of resting or of moving uniformly straight ahead.*
>
> *On Motion*: Definition 2.
> *Moreover, [I call] the force of a body, or the [force] innate in a body, that by which it endeavors to persevere in its own motion along a straight line.*

> *Principia*: Definition 4. *An impressed force is the action exerted on a body in order to change its state of either rest or motion uniformly straight ahead.*

> *Principia*: Definition 5.
> *A centripetal force is that whereby a body is drawn, impelled or in any manner tends toward some point, its center, so to speak.*
>
> *On Motion*: Definition 1.
> *I call centripetal the force by which a body is impelled or attracted toward some point which is regarded as the center.*

The following three definitions further refine Newton's view of a centripetal force by listing three specific quantities: the "absolute quantity," the "accelerative quantity," and the "motive quantity."

> *Principia*: Definition 6. *The absolute quantity of centripetal force is the measure of this force that is greater or less in proportion to the efficacy of the cause propagating it from the center through the surrounding regions.*

Newton relates the absolute quantity of the force to "the center of force" itself, such as "the earth for the center of gravitational force." The more massive the earth, the greater the centripetal force that it will exert upon a unit mass.

> *Principia*: Definition 7. *The accelerative quantity of centripetal force is the measure of this force that is proportional to the velocity that it generates in a given time.*

Newton relates the accelerative quantity of the force to "the place of the body," such as "the gravitating force is greater in valleys and less on the peaks of high mountains." The more distant the center of force, the lesser the centripetal force that acts on a unit mass.

> *Principia*: Definition 8. *The motive quantity of centripetal force is the measure of this force proportional to the motion that it generates in a given time.*

Newton relates the motive quantity of the force to "the body itself," such as the "propensity of the whole toward the center compounded of the propensities of all the parts." The motive quantity of a centripetal force is measured by the change in the quantity of motion (the change in the product of the velocity and quantity of matter) that it generates in a given time in a given mass. Newton stated that "for the sake of brevity, we may call these quantities of forces absolute, accelerative, and motive forces." It is important to note that the motive force depends upon the time required to generate the change in motion. When Newton gave his second law of motion, he stated that "a change in motion is proportional to the motive force impressed." If Law 2 implies that the change must be in a "given time," then Definition 8 confirms that implication. It has been argued, however, that Newton's concept of time as particulate permits quite a different reading of Law 2 (see the discussion on page 146).[2]

> *Principia*: Scholium. *Thus far it has seemed best to explain in what sense less familiar words should be taken in what follows. For time, space, place, and motion I do not define since they are very well known to all. But I should say that common opinion conceives of these quantities only in relation to perceivable objects. And hence certain preconceptions arise, and in order for these to be removed, it is useful to separate them into absolute and relative, true and apparent, and mathematical and common* [*opinions*].
>
> 1. *Absolute, true, and mathematical time, in itself and by its own nature, flows...*
> 2. *Absolute space by its own nature and without reference to anything external...*
> 3. *Place is the part of space which a body occupies, and it is either absolute or relative...*
> 4. *Absolute motion is the translation of a body from an absolute place...*

After some ten pages of detailed discussion of these topics, Newton concludes the scholium with the following statement:

> *In the following pages, however, it will be set forth more fully how to determine true motions from their causes, effects, and apparent differences, and, conversely, how to determine from motions, whether true or apparent, their causes and effects. To this end I have composed the following treatise.*

This long scholium has deservedly occupied the attention of many Newtonian scholars. It is not reproduced in its entirety here: in part because it already has been the object of much attention, and in part because the goal of my work is to detail Newton's analysis, which I believe undergoes relatively little change, even if the concepts employed in it undergo a gradual clarification. The topics of time and space, however, are so central to all of Newton's analyses, and to the worldview of those who follow, that a few comments on its role are in order.[3]

The key word is *absolute*: absolute time, absolute space, and absolute motion. The scholium represents Newton's attempt to explain *visible* relative motion by relating it to *invisible* absolute space and time. Attempts to understand the concepts of time and space can be found as early as the writings of Plato and Aristotle. For Plato, "time is a moving image of eternity," which refers to the eternal uniform circular motion of celestial bodies.[4] Aristotle amended Plato's view by identifying time with "the numerical aspect" of the motion of these material bodies, rather than with the motion itself.[5] Aristotle then defined space in terms of a material continuum and the place in such a space as "the outer limit of that contained."[6] In contrast to this Aristotelian concept of motion as fixed to visible matter, Newton's concept of motion is fixed to invisible space. Newton's view of absolute space and time ultimately was rejected by physics in the twentieth century, but it provided a basis for physics in the eighteenth and nineteenth centuries.

Newton built upon Galileo's analysis of motion. In support of the Copernican system, Galileo had to overcome the common belief that if the earth moved, then falling bodies would be left behind. He did so by arguing that one cannot demonstrate absolute rectilinear motion.[7] In Corollary 5 to the first law of motion (see the following), Newton states that the motion of bodies "are the same with respect to each other whether that space is at rest or moves uniformly in a straight line." Thus, it is impossible to construct an experiment to determine an absolute state of rest (though Newton will attempt to distinguish between a state of rotation and a state of rest in a discussion of the physics of a rotating bucket of water). Nevertheless, he postulates a space at absolute rest, perhaps with respect to the "fixed stars." Newton may have been motivated in part by his desire to refute Descartes's relativism of motion (and the complete denial of space by the Cartesian continuum of matter).[8] In any event, Newton sets forth in this scholium the view of space and time that supports his analysis of motion and that became the accepted, "obvious" basis for the very successful science of the eighteenth and nineteenth centuries. Only after extended analysis of the invariance of the laws of physics in the twentieth century did Einstein and others have reason to challenge Newton's received view of absolute space and time.[9] Julian Barbour neatly sums up the contribution of Newton's view to the development of dynamics:

> The successes of the *scholium* speak for themselves. Within half a century or so Newtonian dynamics conquered the world. Men came to accept his concepts of absolute space and time—and they worked brilliantly. Descartes's confused notions were completely forgotten. You had to look quite hard to find the faults in the *scholium*.[10]

It is interesting that the change that comes about in the twentieth century—which requires the invariance of the laws of motion to be primary to any preconceived notion of absolute space and time—gives rise to an Einsteinian concept of space and time that is closer to the Aristotelian concept of motion (as fixed to visible matter) than to the Newtonian concept of motion (as fixed to invisible space).

AXIOMS OR LAWS OF MOTION

Following the opening list of definitions in both *On Motion* and the *Principia*, Newton presented a set of "Axioms or Laws of Motion." Some of these items appear under different titles in *On Motion*, but some were added for clarification.

> *Principia*: Law 1.
> Every body perseveres in its state of resting or of moving uniformly straight ahead except insofar as it is compelled by impressed forces to change that state.

> *On Motion*: Hypothesis 2.
> Every body by its innate force alone progresses uniformly along a straight line to infinity unless something impedes it from outside.

> *Principia*: Law 2. A change in motion is proportional to the motive force impressed, and takes place along a straight line on which that force is impressed.

See Definition 8 in which the "motive quantity of centripetal force" is, "for the sake of brevity," simply to be called the "motive force." Thus, the motive force is proportional to the change in the quantity of motion (the change in the product of the velocity and the quantity of matter) that it generates in a given time. Importantly, the element of time is *explicit* in this definition of motive force and hence the element of time is *implicit* in Law 2. Bernard Cohen has argued, however, that Newton conceived the element of time to be particulate, even though his concept of time was continuous. Thus, Law 2 stands on its own for Newton if seen in terms of impulse $I = \Delta mv$ rather than in terms of force $F = \Delta mv/\Delta t$, because the element of time Δt as a constant is not relevant in ratios or proportions.[11]

> *Principia*: Law 3. To any action there is always an opposite and equal reaction; in other words, the actions of two bodies upon each other are always equal and are opposite in direction.

> *Principia*: Corollary 1.
> A body, with forces having been conjoined, describes the diagonal of a parallelogram in the same time as it describes the sides, with [forces] having been separated.

> *On Motion*: Hypothesis 3.
> A body, in a given time, with forces having been conjoined, is carried to the place where it is carried by separated forces in successively equal times.

Note that the measure of a force for Newton is always the displacement it produces in a given time. Hence, the parallel rule above applies to displacements: "is carried to the place" or "describes the sides."

> *Principia*: Corollary 2. *Obvious from this is the compounding of the direct force* AD *from any oblique forces* AB *and* BD *and, conversely, the resolving of any direct force* AD *into any oblique ones* AB *and* BD *whatsoever. The validity of this composition and resolution is, indeed, abundantly confirmed from mechanics.*

Again, Newton intends the "compounding" or "resolving" of any "direct force" to mean the addition or resolutions of displacements produced by the force.

> *Principia*: Corollary 3. *The quantity of motion, which is determined by adding the motions made in one direction and subtracting the motions made in the opposite direction, is not changed by the action of bodies on one another.*

> *Principia*: Corollary 4. *The common center of gravity of bodies does not by their interactions change its state of motion or of rest, and consequently the common center of gravity of all bodies acting mutually upon each other (excluding external actions and impediments) either is at rest or moves uniformly straight forward.*

> *Principia*: Corollary 5. *When bodies are confined in a given space, their motions are the same with respect to each other whether that space is at rest or moves uniformly in a straight line without any circular motion.*

> *Principia*: Corollary 6. *If bodies move in any manner whatsoever with respect to each other and are urged on in parallel directions by equal accelerative forces, they will all proceed to move in the same manner with respect to each other, as if they had not been acted upon by those forces.*

> *Principia*: Scholium. *Thus far I have set forth principles accepted by mathematicians and confirmed by a multiplicity of experiments. By means of the first two laws and their first two corollaries Galileo discovered that the descent of heavy bodies is in the doubled ratio of the time, and that the motion of projectiles takes place in a parabola, in agreement with experience, except insofar as those motions are slowed somewhat by the resistance of the air. From the same laws and corollaries depend what has been demonstrated regarding the periods of vibrating pendulums, supported by our daily experience of clocks. From these same ones again, along with the third law, Mr. Christopher Wren, Dr. John Wallis, and Mr. Christiaan Huygens—easily the principal geometers of the present age—separately derived rules for the collision and recoil of two bodies, communicating them almost about the same time to the Royal Society in forms exactly (in regard to these rules) in agreement with each other. And Wallis was indeed the first to publish what had been discovered, and then Wren and Huygens. But it was Wren who confirmed the truth of these [rules] publicly before the Royal Society by means of the experiment with pendulums, which the renowned Mariotte also thought worthy of an entire book soon afterward.*

In this scholium Newton also gave the reason for setting out the following group of eleven lemmas, which he does before beginning his analysis of any particular problem:

> [Scholium] *Indeed I have placed these Lemmas first in order to avoid the tedium of deducing complicated proofs 'ad absurdum' in the manner of ancient geometers. For proofs are rendered more compact by the method of indivisibles. . . . Hence in what follows, . . . if for straight line segments I use minute curved ones, [then] I want it understood . . . that the force of proofs must always be referred to the method of the preceding lemmas.*

The final section of Newton's introductory material concludes with a long essay on experiments.

SECTION 1

Section 1. *On the Method of first and last Ratios, with the help of which the following [lemmas] are proved.*

This set of lemmas provides a formal retrospective defense of the use of the limiting procedures that Newton had employed, with no formal defense, in the tract *On Motion*. His intention was to use these lemmas to provide a formal defense for the limiting procedures employed in the *Principia*. In the body of the propositions that follow in Sections 2 and 3, however, Newton rarely explicitly called upon the opening set of eleven lemmas that make up Section 1. As the editor of Newton's mathematical papers put it,

> It would appear that his initial vision of presenting a logically tight exposition of the principles of motion under accelerative forces faded more and more when he came in detail to cast his arguments, and that he was happy after a while to lapse into the less rigorously justified mode of presentation which he largely exhibits in his published *Principia*. Once into the body of the proofs in the *Principia*, Newton reverted to the more intuitive mode of expression that he used in the tract *On Motion* (e.g., limiting processes, such as those in which an element of arc is reduced and is replaced by its chord). Newton employs these limiting processes without a specific formal defense, such as a reference to this initial set of lemmas.[12]

In what follows, I give abbreviated statements of the lemmas without their figures or justification in order to give an overview of Newton's argument. The full statements and demonstrations are given with figures in the translation in the Appendix. The exception is Lemma 11, which is discussed in some detail in this chapter.

> *Principia*: Lemma 1. *Quantities, as well as ratios of quantities, that constantly tend to equality in a given time, and in that way are able to approach each other more closely than for any given difference, come ultimately to be equal.*

In contemporary terms, this lemma states that the "limit of a difference is the difference of the limits." Symbolically, if the limit of the difference $[f(x) - g(x)]$ goes to zero as x goes to a given value a, then the limit of $f(x)$, as x goes to a, is equal to the limit of $g(x)$, as x goes to a. This lemma is used, for example, in Case 3 of Proposition 9 to argue that if two angles are "constructed by this law common to both and approach nearer to each other than for any assigned difference," then "by Lemma 1, [they] will ultimately be equal."

Principia: Lemma 2. *If in any figure . . . there should be inscribed any number of parallelograms . . . and then the width of these parallelograms should be diminished, and the number should be increased indefinitely; I assert that the ultimate ratios that the inscribed figure . . . , the circumscribed [figures] . . . , and the curvilinear [figure] . . . , have to each other are ratios of equality.*

Principia: Lemma 3. *The same last ratios are also of equality when sides . . . of the parallelograms are unequal, and all are diminished indefinitely.*

Principia: Lemma 4. *If in two figures . . . there should be inscribed (as above) two series of parallelograms, and if the number of both should be the same, and when the widths are diminished indefinitely, and the last ratios of parallelograms in one figure should be individually the same as the parallelograms in the other figure; [then] I assert that the two figures . . . are in the same ratio to one another.*

Principia: Lemma 5. *[In] similar figures, all sides which correspond to each other mutually, curvilinear as well as rectilinear, are proportional, and their areas are in the doubled ratio of their sides.*

Principia: Lemma 6. *If any arc AB given in position should be subtended by the chord AB, and [if] at some point A in the middle of its continuous curvature, it should be touched by the straight line AD extended in either direction; then [if] points A and B should approach each other and coalesce; I assert that the angle BAD [generated] by the chord and tangent, would be diminished indefinitely and would ultimately vanish.*

Principia: Lemma 7. *With the same suppositions, I assert that the last ratio of the arc, chord, and tangent to each other is the ratio of equality.*

Principia: Lemma 8. *If the given straight lines AR and BR constitute with the arc AB, its chord AB and tangent AD the three triangles ARb, ARB, and ARD; and then [if] the points A and B approach each other; I assert that the ultimate forms of the vanishing triangles is one of similarity, and the last ratio, of equality.*

Principia: Lemma 9. *If straight line AE and curve AC given in position mutually intersect at the given angle A, and [if] BD and EC should be applied as ordinates to that straight line at any given angle, meeting the curve at B and C, and if the points B and C approach point A; [then] I assert that the areas of triangles ADB and AEC will be ultimately to each other in the doubled ratio of the sides.*

150   THE REVISIONS AND EXTENSIONS

Principia: Lemma 10.
Spaces which a body describes at the urging of any standard force are, at the very beginning of motion, in the doubled ratio of the times.

On Motion: Hypothesis 4.
Spaces which a body describes at the urging of any centripetal force are, at the very beginning of motion, in the doubled ratio of the times.

The opening statements of Lemma 10 and Hypothesis 4 are identical except that the more restrictive "centripetal force" has been changed to the unrestricted "standard force." In the first draft of *On Motion*, no text accompanies the statement of Hypothesis 4. In the second draft the hypothesis becomes a lemma and Newton provides a demonstration.

LEMMA 11: A DETAILED DISCUSSION

Principia: Lemma 11. *The vanishing subtense of an angle of contact [the line BD] is ultimately in the doubled ratio of the subtense of the conterminous arc [the line AB].*

Figure 7.1 is based upon the diagram that appears in the *Principia*. The lemma demonstrates that the line BD is proportional to the square of the line AB as the point B approaches the point A. In the 1687 edition of the *Principia*, Newton employs this result in the scholium to Proposition 4 on circular motion and in Proposition 9 on spiral motion. In the revised editions of the *Principia*, however, this result also plays a primary role in Proposition 6, in which the basic linear dynamics ratio is derived. The demonstration of this lemma makes reference to "the properties of circles," by which Newton means the existence of a unique circle of curvature at the general point A on the curve. Newton makes this reference clearer in the revised editions of the *Principia* when he inserts in the opening statement of the lemma the qualification, "in all curves which have finite curvature at the point of contact." In the 1687 edition, however, the reader has no advanced warning that the concept of curvature lies behind Newton's demonstration. The opening statement of the lemma is followed by three cases, which I discuss in detail.

Case 1. *Let AB be the arc, AD its tangent, BD the subtense of the angle of contact perpendicular to the tangent, and AB the subtense of the arc. Perpendicular to the latter subtense AB and to the tangent AD, erect AG and BG, meeting at G; then let the points D, B, and G approach the points d, b, and g;*

Here, the elements of the diagram in figure 7.1 are defined. A subtense is simply the chord that subtends a given angle or arc. The statement of Case 1 continues.

[Case 1] *and let J be the intersection of the lines BG and AG, ultimately occurring when the points D and B approach up to A. It is obvious that the distance GJ can be less than any assigned one.*

THE *PRINCIPIA* AND ITS RELATIONSHIP TO *ON MOTION*   151

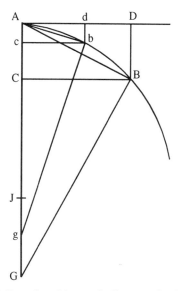

Figure 7.1 Based on Newton's diagram for Lemma 11.

Figure 7.2 is the diagram based on Newton's original drawing into which has been added a set of three circles drawn tangent to the curve at point A. The two outer circles pass through the points B and b. The innermost circle represents the circle of curvature to the curve at point A and the line AJ is the given finite diameter of curvature. As the point B approaches the point A, the outer circles approach the circle of curvature and the diameters of those circles approach the diameter AJ. Thus, the point G approaches the point J in the limit as the point B approaches the point A and, as Newton states, "It is obvious that the distance GJ can be less than any assigned one."

[Case 1] *Now (from the nature of circles passing through points* ABG *and* Abg) AB² *is equal to* AG x BD,

Critical to the demonstration of this lemma is the parenthetical expression: "(from the nature of circles . . .)." As the point B approaches the point A along the general curve AbB, then the set of circles drawn through the points ABG and Abg will approach the limiting circle of curvature drawn through the points A and J. This parenthetical expression is the only explicit reference to the underlying concept of curvature. Figure 7.3 gives the details of the circle drawn through the points A, B, and G. The line AC is perpendicular to the line BC by construction. The line AB is perpendicular to the line BG because AB and BG are chords of the circle ABG and hence, from Euclid's *Elements*, Book 3, Proposition 31, the angle between

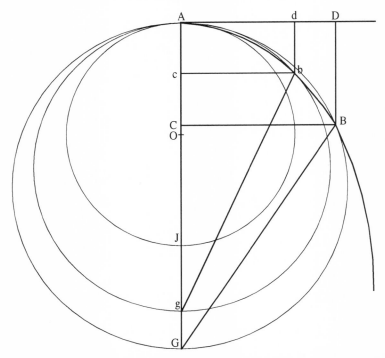

Figure 7.2 A revision of Newton's diagram for Lemma 11 with a set of three circles drawn tangent to the curve at point A. The innermost circle AJ represents the circle of curvature at point A.

chords drawn to a diameter is a right angle. Thus, triangles ABC and AGB are similar and $AG/AB = AB/AC$, where $AC = BD$ by construction. Thus, $AB^2 = AG \times BD$ as given above.

> [Case 1] *and* Ab^2 *is equal to* $Ag \times bd$, *and hence the ratio* AB^2 *to* Ab^2 *is composed of the ratios* AG *to* Ag *and* BD *to* bd.

By a similar demonstration to that given above, $Ab^2 = (Ag)(bd)$ and the ratio of $AB^2/Ab^2 = (AG)(BD)/(Ag)(bd) = (AG/Ag)(BD/bd)$ or "the ratio AB^2 to Ab^2 is composed of the ratios AG to Ag and BD to bd." Case 1 concludes with the following statement:

> [Case 1] *But since* JG *can be assumed less than any assigned length, it can be arranged that the ratio* AG *to* Ag *differs from the ratio of equality less than for any assigned difference, and thus as the ratio* AB^2 *to* Ab^2 *differs from the ratio* BD *to* bd *less than for any assigned difference. By Lemma 1 there is, therefore, the last ratio* AB^2 *to* Ab^2 *equal to the last ratio* BD *to* bd. *Which was to be proven.*

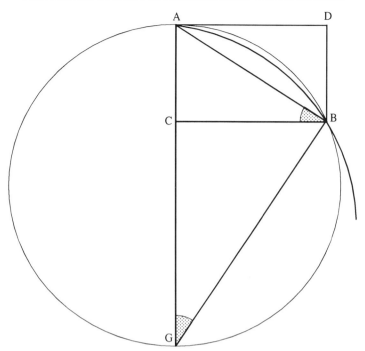

Figure 7.3 Details of the circle *ABG* from figure 7.2. The triangles *ABC* and *AGB* are similar.

Since both of the diameters *AG* and *Ag* approach the given finite diameter *AJ* of the circle of curvature as the points *B* and *b* approach the point *A*, then the ratio *AG* / *Ab* approaches unity and the "last ratio" AB^2 / Ab^2 becomes equal to the "last ratio" *BD* / *bd*.

> Case 2. *Now let* BD *be inclined to* AD *at any given angle, and there will always be the same last ratio as before and hence also the same* AB^2 *to* Ab^2, *as was to be proven.*

If the line *BD* were not perpendicular to the tangent line *AD*, then the perpendicular *BD'* would be *BD'* = *BD* cosθ, where θ is the given angle. The angle would be the same for *bd* as for *BD* and hence the ratio *BD'*/*bd'* = *BD* cosθ / *bd* cosθ = *BD*/*bd* and thus the last ratio will be unchanged.

> Case 3. *Moreover, although the angle* D *may not be given,* [*if* BD *converge to a given point, or be drawn according to any other law*] *the angles* D *and* d [*constructed by this law common to both*] *will always verge toward equality and approach nearer to each other than for any assigned difference, and hence, by Lemma 1, will ultimately be equal, and accordingly, the lines* BD, bd *will be in the same ratio to each other as before. Which was to be proven.*

In the revised editions of the *Principia*, Newton enlarged this statement by inserting the qualifications given in brackets. Thus, if the inclination of *BD* is determined by its relationship to a given curve, as with the circle in Proposition 4 and the spiral in Proposition 9, then the last ratio remains the same and the relationship given in the lemma holds.

SECTION 2

Section 2. *Of the invention of centripetal forces*

In the first theorem in the tract *On Motion*, Newton derived Kepler's law of equal areas in equal times. That same demonstration, with minimal revisions, appears as Proposition 1 of the 1687 *Principia*. In the second theorem in *On Motion*, Newton discussed uniform circular motion and that topic is found in Proposition 4 of the 1687 *Principia*. The third theorem in *On Motion* contains the general paradigm for the solution of direct problems and that demonstration, with minimal revisions, appears as Proposition 6 in the 1687 *Principia*. Propositions 7 to 10 in the 1687 *Principia* give examples of how to apply that paradigm to specific orbits and centers of force, as did Problems 1 and 2 in *On Motion*.

Principia: Proposition 1. Theorem 1.	*On Motion*: Theorem 1.
The areas that bodies driven in orbits describe with radii having been constructed to a stationary center of force, lie in stationary planes and are proportional to the times.	*All orbiting bodies describe, by radii having been constructed to their center, areas proportional to the times.*

Newton shifted from the active statement "orbiting bodies describe" in *On Motion* to the passive statement "bodies driven in orbit describe" in the *Principia*. He will make the same change in the problems to follow, where "a body orbits" becomes "let a body be orbited," and so on. The active expression is neutral, whereas the passive may imply an external mover of some sort. In Definition 8, however, Newton states that he considers "forces not physically, but only mathematically. Hence, let the reader beware lest because of words of this kind [attraction or impulse], they should think either that I am defining . . . a physical cause or reason, or that I am attributing forces truly and physically to centers (which are mathematical points)." Clearly, however, Newton considers something active, either internal or external, to be operating on the body in order to produce the given orbit.[13]

The revised opening statement of the *Principia* makes explicit that the center of force is stationary and that the areas lie in fixed (stationary) planes. Otherwise the text is little changed from the text in *On Motion*. The first seven sentences of the *Principia*, which carry one from the opening statement to the climactic statement, "in equal times, therefore, equal areas are described," are identical word for word with the text in *On Motion* except for the inclusion of two parenthetical expressions, "(by Law I)" and "(by Corollary 1 of the Laws)." Newton revises the remainder of the text in order more carefully to qualify the result given in Theorem 1. The reader should refer to the detailed discussion of Theorem 1 in chapter 4 if questions arise in the reading of Proposition 1 in the 1687 *Principia*, which is given in the Appendix.

> *Principia*: Proposition 2. Theorem 2. *Every body that, when it is moved along some curved line with a radius having been constructed to a point [that is] either stationary or advancing uniformly in a rectilinear motion, describes areas proportional to the times around that point, is urged on by a centripetal force being directed to the same point.*

In Proposition 1 it was demonstrated that if a body is driven by a centripetal force, then the radius to the center of force sweeps out equal areas in equal times. In Proposition 2 it is the inverse theorem that is demonstrated: if the radius to the center of force sweeps out equal areas in equal times, then the body is driven by a centripetal force. This theorem does not appear in *On Motion*, but the type of analysis is similar to that employed in the detailed discussion of Theorem 1. One begins with equal triangles, and thus the deviation must be parallel to the line of force, and thus the force is centripetal.[14]

> *Principia*: Proposition 3. Theorem 3. *Every body that, with a radius having been constructed to the center of another arbitrarily moving body, describes areas proportional to the times around that center, is urged by a force compounded of the centripetal force being directed toward the other body, and of the whole accelerative force by which the other body is urged.*

A planetary satellite, such as the moon, moves under the influence of two forces: a force directed toward the planet and a force directed toward the sun. The satellite is the "body" and the planet is the "other body." Proposition 3 extends the results from Proposition 2, with a fixed center of force, to the case of a moving center of force. This theorem does not appear in *On Motion*.

Principia: Proposition 4. Theorem 4.	*On Motion*: Theorem 2.
For bodies that describe different circles with equable motion, the centripetal forces are directed toward the centers of those circles, and are to one another as the squares of the arcs described in the same time divided by the radii of the circles.	*For bodies orbiting uniformly on the circumferences of circles, the centripetal forces are as the squares of the arcs described in the same time divided by the radii of the circles.*

The expanded version of the opening statement in the *Principia* notes that the centripetal forces are directed toward the center of the circles, a point only implicit in the earlier tract *On Motion*. The first halves of the texts are identical, word for word, with the exception of the phrase "distances *CD*" which becomes the "nascent intervals *CD*." The second half of the revised text is expanded by reference to Theorem 2, Lemma 5, and Lemma 11. The reader who has worked through the detailed discussion of Theorem 2 in chapter 4 will be able to work through the revised demonstration of Proposition 4, which is given in the Appendix.

In the scholium to Proposition 4, Newton states that he has "decided to explain more fully" questions concerning the circular centripetal force. Following the opening demonstration of Proposition 4 in this scholium is a revised version of his pre-1669 discussion in the *Waste Book* of uniform circular motion using the polygonal approximation. Newton told Halley how he came across the earlier version "in turning over some old papers."[15] It has also been suggested "that Newton added this final paragraph to the scholium as a means of asserting his proper priority over Hooke."[16] In the pre-1669 version, Newton did not determine the nature of the force, but in the revised version for 1687 he does. Both versions have been discussed in detail in chapter 3.

Principia: Proposition 5. Problem 1. *Given, in any places, a velocity by which a body describes a given figure by forces directed to any common center, to find that center.*

Newton added this proposition just before the manuscript was sent to press. It has utility in the analysis of celestial observations in which the location of the center of force can be found from observations of the movement of a body, such as a comet. As such, it does not play a role in the analysis of direct problems that are of central interest here.

Principia: Proposition 6. Theorem 5. *If a body* P *by revolving around the center* S, *should describe any curved line* APQ, *and if the straight line* ZPR *should touch that curve at any point* P, *and if to this tangent from any other point* Q *of the curve,* QR *should be drawn parallel to the distance* SP, *and if* QT *should be dropped perpendicular to the distance* SP; [*then*] *I assert that the centripetal force would be reciprocally as the solid* $SP^2 \times QT^2 / QR$, *provided that the quantity of that solid that ultimately occurs when the points* P *and* Q *coalesce is always taken.*

The text of Proposition 6 in the *Principia* is identical with the text of Theorem 3 in *On Motion* with the exception of the following revisions. In the opening statement of the theorem "orbiting" (gyrando) is replaced by "revolving" (revolvendo) and the tangent is described as the "straight line *ZPR*" rather than the "straight line *PR*." The first halves of the texts are identical, word for word, except for the insertion of "nascent" to modify the "line segment *QR*" and the inclusion of two parenthetical expressions, "(by Law 2)" and "(by Lemma 10)." The second half of the text is slightly revised to make explicit the substitution of the area $SP \times QT$ for the time in the expression for the force. The reader should refer to the detailed discussion of Theorem 3 in chapter 4 if there are questions concerning Newton's demonstration of Proposition 6 of the 1687 *Principia*, which is given in the Appendix.

Principia: Proposition 7. Problem 2. *Let a body be orbited on the circumference of a circle; there is required the law of centripetal force being directed to some given point on the circumference.*

On Motion: Problem 1. *A body orbits on the circumference of a circle; there is required the law of centripetal force being directed to some point on the circumference.*

The opening statements are essentially unchanged, with the exception of the active "a body orbits" becoming the passive "let a body be orbited" (see the discussion of Proposition 1). Otherwise, the bodies of the texts are identical, word for word, except for three clarifications: (1) the "line SP" is now described as the "straight line SP," (2) the parenthetical expression "(by the Corollary of Theorem 5)" is inserted, and (3) an explicit reference is made to the similarity of triangles that was only implicit in *On Motion*. (In fact, the figure is slightly revised by the addition of the point Z to clarify this relationship.) The reader should refer to the detailed discussion of Problem 1 in chapter 4 if there are questions concerning Proposition 7 of the 1687 *Principia*.

Principia: Proposition 8. Problem 3. *Let a body be moved on a circle PQA: for this effect there is required the law of centripetal force being directed to a point at such a distance that all lines* PS *and* RS *constructed to it can be considered as parallels.*

This solution is an extension of the problem considered in the previous proposition in which the force center was removed an indefinitely large distance. The scholium claims that the solution can also be applied to other conics but, according to Whiteside, that generalization is invalid.[17]

PROPOSITION 9: A DETAILED DISCUSSION

Principia: Proposition 9. Problem 4. *Let a body be orbited on a spiral* PQS *intersecting all the radii* SP, SQ, *etc. at a given angle; there is required the law of centripetal force being directed to the center of the spiral.*

In the scholium to Problem 1 in the tract *On Motion*, Newton simply referred to the solution to this problem, "In a spiral which cuts all the radii at a given angle, the centripetal force being directed to the beginning of the spiral is reciprocally in the tripled ratio of the distance." He did not give a demonstration in *On Motion*, but he does give one in the *Principia*. I give the statement of the demonstration next and then follow it with a line-by-line analysis. Read the full statement to get an overview of the proposition before attempting to justify it, and then follow the details in the

158   THE REVISIONS AND EXTENSIONS

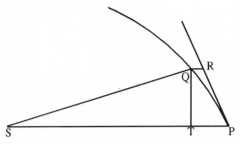

Figure 7.4 The diagram for Proposition 9. The line *PQ* represents a spiral with its pole at the center of force *S*.

line-by-line analysis. Figure 7.4 is similar to the drawing that appears in the *Principia*.[18]

Demonstration

> Let there be given the indefinitely small angle PSQ, and, because all the angles have been given, the figure SPQRT will be given in species. The ratio QT / RQ is therefore given, and there is QT² / QR as QT, that is, as SP.
>
> Now let the angle PSQ be changed in whatever manner, and the straight line QR subtending the angle of contact QPR will be changed (by Lemma 11) in the doubled ratio of PR or QT. Therefore QT² / QR will remain the same as before, that is, as SP.
>
> For this reason, QT² x SP² / QR is as SP³, that is (by the Corollary of Theorem 5) the centripetal force is [reciprocally] as the cube of the distance SP. Which was to be done.

The demonstration is very brief and Newton appears to employ two properties of the spiral without explicitly stating them. The first property is that the angle of contact between the tangent *PR* and the radial distance *SP* is a constant (i.e., the spiral in question is an "equal angle spiral").[19] The second property is that the radius of curvature of the arc *PQ* is proportional to the radial distance of the pole *SP*.[20] Newton did not need to employ the concept of curvature directly in solutions to the problems he elected to demonstrate in the tract *On Motion*. Here in the *Principia*, however, curvature must be employed in the solution of this spiral/pole direct problem. In the following, the demonstration of the proposition is considered in detail.

[A]   *Let there be given the indefinitely small angle* PSQ, *and, because all the angles have been given, the figure* SPQRT *will be given in species.*

The demonstration begins with the assumption that angle *PSQ* be indefinitely small and that it initially be given. Later in the demonstration the

THE *PRINCIPIA* AND ITS RELATIONSHIP TO *ON MOTION*

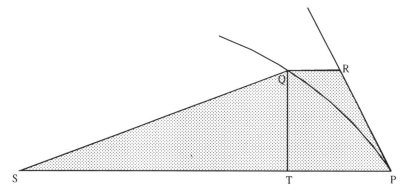

Figure 7.5 The shaded area is the figure *SPQRT* that is "given in species."

angle *PSQ* will be permitted to change but here as the proof begins it is fixed. Thus, (1) given the angle *PSQ*, (2) given the construction of *QR* parallel to *SP*, and (3) given the property of the equiangular spiral that the angle *SPR* is a constant, then all the angles in the figure *SPQRT* are given. The shaded area in figure 7.5 is the figure *SPQRT*. In general, the specification of all the angles of any figure is not sufficient to assure that the figure is "given in species" (i.e., the ratios of the sides of the figure are in a given ratio). The specification of all angles in a triangle would be sufficient, for example, to fix the ratio of its sides, but it would not be sufficient for a rectangle. Note, however, that Newton has specified that the given angle *PSQ* is "infinitely small." In that small limit, the arc of any curve in the vicinity of the point *P* can be represented by the arc of the circle of curvature at that point. The points *P* and *Q*, therefore, lie on the circumference of the circle of curvature and for a given small angle *PSQ* the sides *QR*, *QS*, *RP*, and *SP* of the figure *SPQRT* are proportional to the radius *PC* of that circle of curvature (i.e., the ratios *QR* / *PC*, *QS* / *PC*, *RP* / *PC*, *SP* / *PC* are given).[21]

That statement holds in general for any curve and for any such figure. The figure in question, however, is for an equiangular spiral where the point *S* is the pole of that spiral. The equiangular spiral has the special property that the pole distance *SP* is in a definite proportion to the radius of curvature *PC*, and hence each element in the figure is in a definite proportion to *SP*. Thus, the ratio of the sides of the figure is determined.[22] To display that relationship more clearly, the spiral *PQS* in figure 7.6 is extended into its pole *S* and the circle of curvature *PVD* is displayed. The chord *PV* of the circle of curvature drawn through the pole of the spiral is bisected by that pole.[23] Therefore, the angle *SCP* is equal to the constant angle *SPR* (since *RPC* = *PSC* = 90°) and thus *SP* = *PC* sin*SPR*. Now, the side

160   THE REVISIONS AND EXTENSIONS

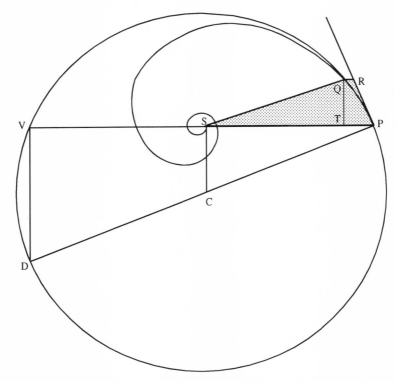

Figure 7.6 The diagram for Proposition 9 with the spiral *PQS* extended into its pole *S* and the circle of curvature *PVD* displayed.

SP is given in addition to all the angles, and thus all the sides and all the ratios of the sides are given (for a given small angle *PSQ*). As Proposition 9 states, therefore, for a given "small angle *PSQ*," the quadrilateral figure *SPQRT* for an equiangular spiral is given "in species" (i.e., the ratios of the sides of the figure are in a given ratio).[24]

   [B] *The ratio* QT / RQ *is therefore given,*

From [A], for a given small angle *PSQ* the ratios of all the elements in the figure *SPQRT* are given.

   [C] *and there is* QT² / QR *as* QT, *that is, as* SP.

For a given small angle *PSQ*, the product of ratios $(QT / QR)(QT / SP)$ is also given or what is the same, the ratio (QT^2 / QR) is proportional to *SP*.

   [D] *Now let the angle* PSQ *be changed in whatever manner, and the straight line* QR *subtending the angle of contact* QPR *will be changed (by Lemma 11) in the doubled ratio of* PR *or* QT.

Now the angle *PSQ* remains indefinitely small but it is permitted to change. Lemma 11 establishes that the ratio QT^2 / QR approaches a finite limit as the point *Q* approaches the point *P* (i.e., as the angle *PSQ* becomes "indefinitely small" as required in the statement of Proposition 9). See the preceding discussion of Lemma 11 in which the ratio is expressed as AB^2 / BD, where (in Case 2) $BD = QR$ and $AB = PQ$ (the chord) or, in the limit, *PR* (the tangent). In the example of an equiangular spiral, $QT = PR \sin SPR$, where *SPR* is the fixed angle of the spiral. Thus, the ratio AB^2 / BD from Lemma 11 is given for Proposition 9 as the ratio QT^2 / QR. For all curves that have finite curvature at the point of contact *P* (a qualification that Newton later inserts in the revised statement of Lemma 11), the ratio QT^2 / QR has a finite limit and hence in that limit *QR* is proportional to QT^2, or as Newton puts it, "the straight line *QR* . . . will be changed (by Lemma 11) in the doubled ratio of *PR* or *QT*."

> [E] *Therefore* QT^2 / QR *will remain the same as before, that is, as* SP. *For this reason,* $QT^2 \times SP^2 / QR$ *is as* SP^3, *that is (by the Corollary of Theorem 5) the centripetal force is [reciprocally] as the cube of the distance* SP. *Which was to be done.*

The force is inversely proportional to the linear dynamics ratio $SP^2(QT^2 / QR)$ and the discriminate ratio (QT^2 / QR) is proportional to *SP*, thus the force is inversely proportional to SP^3, or as Newton states above, "the centripetal force is [reciprocally] as the cube of the distance *SP*. Which was to be done."

LEMMA 12 AND PROPOSITION 10

Following the solution to Proposition 9, Newton inserted Lemma 12, a relationship that appeared as Lemma 1 in *On Motion* and that is required in the analysis of the direct problem that appears in Proposition 10.

> *Principia*: Lemma 12.
> *All parallelograms described around a given ellipse are equal to each other. Understand the same for parallelograms described in a hyperbola around its diameters. This is evident from* Conics.

> *On Motion*: Lemma 1.
> *All parallelograms described around a given ellipse are equal to each other.*
>
> *This is established from the* Conics.

This relationship is demonstrated in Book 7, Proposition 31 in the *Conics* of Apollonius.[25] Newton has added the reference to a hyperbola because in the 1687 *Principia* he discussed hyperbolic orbits as well as elliptical orbits. See the discussion and diagram for Lemma 1 in chapter 4.

> *Principia*: Proposition 10. Problem 5.
> *Let a body be orbited on an ellipse; there is required the law of centripetal force being directed to the center of the ellipse.*

> *On Motion*: Problem 2.
> *A body orbits on a classical ellipse; there is required the law of centripetal force being directed to the center of the ellipse.*

The opening statements of this proposition in *On Motion* and in the *Principia* are almost identical, except for the change from the active, "a body orbits," to the passive, "let a body be orbited." The bodies of the texts differ only by the addition of three parenthetical expressions: "(from the *Conics*)," "(by Lemma 12)," and "(by the Corollary to Theorem 5)," and by the addition of a line of qualification concerning a set of propositions. Except for a slight rephrasing of one other line, the texts are identical, word for word. The reader should refer to the detailed discussion of Problem 2 in chapter 5 if there are questions concerning Proposition 10 of the 1687 *Principia*.

SECTION 3

Section 3. *Of the motion of bodies in eccentric Conic Sections*

The preceding section contains the solutions to the direct problems of a circular path with a center of force on the circumference, a spiral path with the center of force at its pole, and an elliptical path with the center of force at the center of the ellipse. These are preliminary examples of the application of the paradigm of Proposition 6. The direct Kepler problem commands much more respect than do these preliminary examples, however, and it is with Proposition 11 and the solution to that problem that Newton opens this new section. It provides the answer to the question raised by Halley on his visit to Newton, a question that set into motion the activity that eventually resulted in the publication of the *Principia*. As he concluded Proposition 11 Newton referred to "the dignity of the problem" and its place of honor at the beginning of a new section. He also gave the solutions to the other conic sections, the hyperbola and the parabola, as separate propositions rather than in a scholium to the proposition on elliptical motion, as he had done in *On Motion*.

Principia: Proposition 11. Problem 6.	*On Motion*: Problem 3.
Let a body be revolved on an ellipse; there is required the law of centripetal force being directed to a focus of the ellipse.	*A body orbits on an ellipse; there is required the law of centripetal force directed to a focus of the ellipse.*

The opening statements are essentially identical, as are the bodies of the text, except that the active statement "a body orbits" in *On Motion* is now the passive statement "let a body be revolved."[26] Newton has added to the demonstration in the *Principia* the following items: a description of a parallelogram, four parenthetical expressions, a qualification, and references to Lemma 8, Lemma 12, and Theorem 5. He has removed the auxiliary ratio of M to N that appeared in the original solution, and he has replaced the scholium with a closing statement. With these minor exceptions, the texts are identical, word for word, and the reader should refer to the detailed discussion of Problem 3 in chapter 5 if there are questions concerning Proposition 11 of the 1687 *Principia* as it is given in the Appendix.

> *Principia*: Proposition 12. Problem 7. *Let a body be moved on a hyperbola; there is required the law of centripetal force being directed to a focus of the figure.*

Newton did not present a separate demonstration for the solution to the direct problem of hyperbolic orbits in *On Motion*. In Proposition 12 of the *Principia*, however, he has very carefully constructed a demonstration of hyperbolic orbits that follows in detail the demonstration of elliptical orbits in Proposition 11. After the opening lines describing the figure, the construction follows the previous solution line for line with only the few necessary accommodations to the new figure. The reader should refer to the detailed discussion of Problem 3 in chapter 5 if there are questions concerning Proposition 12 of the 1687 *Principia*.

> *Principia*: Lemma 13. *The* latus rectum *of a parabola pertaining to any vertex is quadruple the distance of that vertex from the focus of the figure. This is evident from the* Conics.[27]

> *Principia*: Lemma 14. *A perpendicular dropped from the focus of a parabola to its tangent is a mean proportional between the distance of the focus from the point of contact and the distance from the principal vertex of the figure.*

These demonstrations of the properties of a parabola are required in the analysis of parabolic motion to follow in Proposition 13.[28]

> *Principia*: Proposition 13. Problem 8. *Let a body be moved on the perimeter of a parabola; there is required the law of centripetal force being directed to the focus of this figure.*

Newton did not present a separate demonstration for the solution to the direct problem of parabolic orbits in *On Motion*, but he does in the *Principia*. Of particular interest is the first corollary to this proposition, in which Newton claims that solutions to the three direct problems also provide solutions to the inverse problem.

> *Principia*: Proposition 13. Corollary 1. *From the last three propositions it follows that if any body* P *should depart from position* P *along any straight line* PR, *with any velocity, and is at the same time acted upon by a centripetal force that is reciprocally proportional to the square of the distance from the center, this body will be moved in one of the sections of conics having a focus at the center of forces; and conversely.*

Newton was criticized for failing to defend this assumption and he provided an outline of a defense in an expanded version of this corollary in the 1713 edition of the *Principia*. (The discussion of this point will be continued in chapter 10.) Whether Newton succeeded or failed in providing a satisfactory solution for the inverse problem has been the subject of considerable scholarly debate from the late seventeenth century until the present.[29]

> *Principia*: Proposition 14. Theorem 6. *If several bodies should be revolved around a common center, and the centripetal force should decrease in the doubled ratio of the distances from the center, I say that the* latera recta *of orbits are in the doubled ratio of the areas that bodies describe by radii constructed to the center in the same time.*

This relationship is required in the demonstration of Proposition 15 to follow.

> *Principia*: Proposition 15. Theorem 7. *With the same suppositions, I say that the periodic times in ellipses are in the three-halved ratio of the transverse axes.*
>
> *On Motion*: Theorem 4. *Supposing that the centripetal force is reciprocally proportional to the square of the distance from the center, the squares of the periodic times in ellipses are as the cubes of their transverse axes.*

The revised demonstration of Kepler's "three-halves power law" in the 1687 *Principia*, which is given in the Appendix, is much simpler than the demonstration in *On Motion*, which is discussed in detail in chapter 6.

> *Principia*: Proposition 16. Theorem 8. *With the same suppositions, and with straight lines drawn to bodies that touch the orbits in the same places, and with perpendiculars dropped to these tangents from a common focus, I say that the velocities of the bodies are in a ratio compounded of the ratio of perpendiculars inversely, and the half ratio of the* latera recta *directly.*

The demonstration of this proposition is followed by nine corollaries in which the relationship of the speeds and *latera recta* of conic sections is explored. Specifically, Corollaries 1 and 3 are employed in the proposition to follow in which the nature of a particular type of conic (elliptical, hyperbolic, or parabolic) is given by the relative magnitude of its initial projection speed.

> *Principia*: Proposition 17. Problem 9. *Supposing that the centripetal force be made reciprocally proportional to the square of the distance from its center, and that the absolute quantity of that force is known; there is required* a line *which a body will describe, when released from a given position with a given* velocity *along a given straight line.*
>
> Let the centripetal force directed to point S be that which makes the body p orbit in any given orbit pq. . . . [emphasis added]
>
> *On Motion*: Problem 4. *Supposing that the centripetal force be made reciprocally proportional to the square of the distance from its center, and that the absolute quantity of that force is known; there is required* an ellipse *which a body will describe, when released from a given position with a given* speed *along a given straight line.*
>
> Let the centripetal force directed to point S be that which makes body p orbit in a circle pq. . . . [emphasis added]

The opening statements in Problem 4 from *On Motion* and Proposition 17 from the *Principia* are identical except for a description of the general path as "a line" rather than "an ellipse" and the replacement of the word for "speed" (*celeritate*) with the word for "velocity" (*velocitate*). In the detailed statement of the text, however, Newton has made many changes, including changing the reference "circle" to "any orbit." He has not changed the proof in any substantive way, but clearly he was not satisfied with the presentation in *On Motion*. The interesting changes in the text are from the specific required "ellipse" for body *P* in *On Motion* to the more general required "line" in the *Principia*, and from the "a circle *pq*" for the reference body *p* in *On Motion* to the more general "any given orbit *pq*" in the *Principia*. Neither of these changes are more than cosmetic. In the revision for the *Principia*, Newton soon lets the body *P* "deflect under the compulsion of the centripetal force into the conic section *PQ*." The "conic section" is still more general than a specific "ellipse," but it is still not a general "line." The shift from the reference "a circle *pq*" to "any given orbit *pq*" is simply the recognition that the reference circle is sufficient but not necessary: any orbit will do.[30]

CONCLUSION

Thus, the first three theorems and the first three problems of *On Motion* have provided the basis for the first seventeen propositions of the *Principia*. In Section 1 of Book One, Newton has added detailed definitions, set forth the laws of motion in an axiomatic fashion, and provided a number of lemmas designed to provide a formal background to the limiting procedures that he used in the earlier tract without a defense. Section 2 opens with a demonstration in Proposition 1 of the law of equal areas, which was the first theorem in *On Motion*, and Section 3 closes with the demonstration of Proposition 17, which was also the final problem on planetary motion in *On Motion*. Independent of the additional material, the core of the dynamics of the two works remains essentially unchanged. The basic paradigm for solving problems remains the linear dynamics ratio, which now appears as Proposition 6. Newton follows it with the same two preliminary problems given in *On Motion* plus the addition of a problem on spiral motion. The single method of Proposition 6 is applied to the three preliminary problems and then Newton presents the solution to the distinguished Kepler problem of elliptical/focal motion. In the revisions to follow, however, Newton introduces two other methods to solve the same problems.

EIGHT

Newton's Unpublished Proposed Revisions
Two New Methods Revealed

Following the publication of the first edition of the *Principia* in 1687, Newton began to make corrections in his working copy of the text and to propose revisions and additions for a possible second edition. When, twenty-six years later, in 1713, the second edition was published, many of these hand-written revisions were incorporated. Several revisions, however, never appeared in printed form. Of particular interest are the unpublished revisions of the fundamental dynamics of Sections 2 and 3 of Book One. These revisions, if published, would have provided a dramatically different format for these fundamental sections. They have been masterfully reconstructed by D. T. Whiteside, the editor of Newton's mathematical papers. Whiteside sets out the vision that Newton had of a revision for the 1687 *Principia*:

> Newton came in the early 1690s to conceive a grand scheme of revision of the published *Principia* in which not only its particular verbal and mathematical errors were to be corrected but, much more radically, the redundant in its logical and expository framework was to be cut out and the flimsier portions of the remaining structure were to be strengthened and supported and (where necessary) completely rebuilt.[1]

Newton was not the only person to suggest corrections and revisions to the *Principia* following its publication. A select group of scholars, both in Britain and on the continent, struggled with the work and were eager to note its failures as well as its successes. The Scottish mathematician David Gregory had aspirations (unrequited) of having his notes on the work published in a revised edition or as a separate companion volume. During Gregory's visit to Cambridge in May 1694, Newton showed him the manuscript papers that contained the drafts of the proposed revisions. Whiteside notes that Newton "proved unprecedentedly expansive regarding his intentions, elaborating for him [Gregory] a detailed overview of his plans for revision."[2] In a memorandum written in July 1694, Gregory summarized the revisions that Newton had discussed during his visit. Of particular interest to Newton's dynamics are the opening lines of Gregory's summary:

> Many corrections are made near the beginning: some corollaries are added; the order of the propositions is changed and some of them are omitted and deleted. He [Newton] deduces the computation of the centripetal force of a body tending to the focus of a conic section from that of a centripetal force tending to the center, and this again from that of a constant centripetal force tending to the center of a circle; moreover the proofs given in propositions 7 to 13 inclusive now follow from it just like corollaries.[3]

Gregory went on to list other revisions proposed by Newton for Books Two and Three. It is the fundamental revisions to the opening sections of Book One, however, that command our interest.

Whiteside has called these proposed revisions of the 1690s "radical restructurings." In the 1687 edition of the *Principia*, Newton employed only the linear dynamics ratio as a measure of the force in producing solutions to the direct problems set in Sections 2 and 3 of Book One. In the proposed radical revision, however, he introduced two other related but distinct methods for generating such solutions: the circular dynamics ratio and the comparison theorem.

In the published revised editions of the *Principia* of 1713 and 1726, Newton retains the wording of the statements of the propositions of the 1687 edition with only minimal changes. In dramatic contrast, however, Newton considered major changes in the statements of the propositions in the proposed radical restructurings of the 1690s. He did not attempt in the unpublished radical revisions to conform to the general outline of the 1687 edition as he did in the revised editions that eventually were published in 1713 and 1726.

The statement of the proposed Proposition 1, Kepler's area law, remained in the unpublished revisions as it was in the 1687 edition, although six new corollaries were added. The statements of the next four propositions also remained unchanged. The statement of the proposed Proposition 6, however, underwent a dramatic revision. In the 1687 edition, Proposition 6 introduced the linear dynamics ratio and then was used to generate solutions for a series of direct problems in the following several propositions. The proposed revised Proposition 6 was the first of three new propositions that introduced a new technique for solving direct problems, the comparison theorem, which bore no resemblance to that used in the original Proposition 6. The linear dynamics ratio of the original Proposition 6 was transferred to the proposed Proposition 9, to which

168   THE REVISIONS AND EXTENSIONS

was added yet another measure of force, the circular dynamics ratio. The proposed Proposition 10 provided a measure of force for motion in a conic directed to any point, and the proposed Proposition 11 produced yet another method of attack. In what follows, we look in some detail at the new method of the proposed Propositions 6, 7, and 8, and of the method outlined in the proposed Proposition 11. Only the revisions of the proposed Proposition 9 appear in the published revised editions; they will be discussed in detail in the chapter to follow.

PROPOSED PROPOSITIONS 6, 7, AND 8

An overview of the revisions that Newton considered making in the presentation of his basic dynamics follows. The center column is the location of the proposed propositions in the unpublished revisions that Newton generated after the publication of the 1687 edition (i.e., what Whiteside has called the "radical revisions"). The first and third columns give the location of the material in the 1687 edition and in the revised editions of 1713 and 1726, respectively.

1687 edition	Proposed radical revision	1713 and 1726 editions
Does not appear	New Proposition 6: Similarity Theorem.[4] *Where bodies describe all similar parts of similar figures in proportional times, their centripetal forces are* . . .	Does not appear
Does not appear	New Proposition 7: Proportionality Theorem.[5] *If in two orbits proportional ordinates stand at any given angles on proportional abscissas . . . the centripetal forces will be as* . . . Corollary: *If one of the orbits be a circle and the other orbit any ellipse, and the point S the center of both . . .* [*the force to the center of the ellipse* ∝ PC].	Does not appear
Does not appear	New Proposition 8: Comparison Theorem.[6] Ratio of forces at point P directed to two different centers of force, S and R, for a given common orbit. $F_S : F_R :: (PS \times PR^2) : PT^3$ (where PT passes through S parallel to the tangent at point P and cuts PR at point T).	Proposition 7, Corollary 3

1687 edition	Proposed radical revision	1713 and 1726 editions
Does not appear	Corollary 1. If the orbit is an ellipse and force center S is at the center, then $F_S \propto PS$ and $$F_R \propto PT^3/PR^2.$$	Proposition 7, Corollary 4
Does not appear	Corollary 2. If the orbit is an ellipse and the force center R is at a focus, then the force is proportional to the inverse square of the distance.	Proposition 11, Alternate Method
Does not appear	Corollary 3. If the orbit is a parabola and the force center R is at a focus, then the force is proportional to the inverse square of the distance.	Does not appear
Does not appear	Corollary 4. If the orbit is a hyperbola and the force center R is at a focus, then the force is proportional to the inverse square of the distance.	Proposition 12, Alternate Method
Proposition 6, Body of the text	New Proposition 9. The linear dynamics ratio $QR/QT^2 \times SP^2$ is derived. The circular dynamics ratio $1/SY^2 \times PV$ is derived.[7]	Proposition 6 Corollary 1 Corollary 3
Does not appear	New Lemma 12. An expression is derived for the chords of the circle of curvature to a conic through the center and through a focus.[8]	Proposition 10, Chord through the center
Does not appear	New Proposition 10. The expression for the force at any point in a conic, $F_R \propto PT^3/PR^2$, which was obtained in the new Proposition 8, is now derived from the new Proposition 9 and the new Lemma 12.[9]	Proposition 7, Corollary 4
Proposition 10	Corollary 1. If the center of force is at the center of the conic, then the force is directly proportional to the distance.	Proposition 10
Proposition 11, Proposition 12, Proposition 13	Corollary 2. If the center of force is at a focus of the conic, then the force is proportional to the inverse square of the distance.	Proposition 11, Proposition 12, Proposition 13

1687 edition	Proposed radical revision	1713 and 1726 editions
Proposition 8	Corollary 3. If the center of force recedes to an infinite distance.	Proposition 8
Proposition 7, Body of the text	Corollary 4. If the conic passes into a circle and the center of force is on the circumference, then the force is proportional to the inverse fifth power of the distance.	Proposition 7, Corollary 1

The three specific basic revisions noted by Gregory in his summary statement are readily discernible in the comparison of the initial and proposed propositions:

1. "The computation of the centripetal force of a body tending to the focus of a conic section [is deduced] from that of a centripetal force tending to the center" [Proposed Proposition 8, Corollaries 2, 3, and 4].
2. "This [force toward the center of an ellipse] again from that of a constant centripetal force tending to the center of a circle" [Proposed Proposition 7, Corollary].
3. "The proofs given in Propositions 7 to 13 inclusive now follow from it just like corollaries" [Proposed Proposition 10, Corollaries 1, 2, 3, and 4].

THE THREE TECHNIQUES

As Gregory also noted, Newton has dramatically changed the order of the propositions. The difficulties involved in producing a revised edition so reordered may well be the reason that the "grand scheme of revision" was never carried to completion. In the preface of the 1687 edition, Newton notes, "Some things found out after the rest, I chose to insert in places less suitable, rather than to change the number of the propositions as well as the citations." In the revisions just given, Newton has drastically revised the number, order, and nature of the opening propositions. The extensive revision of citations in the remaining propositions, even those that were not to be changed, would have presented an enormous editorial challenge. In the 1713 edition, Newton chose to continue with his earlier practice of inserting the new material "in places less suitable." In 1694, however, he was still toying with the idea of a dramatic revision. Newton's proposed revision did more than shift his dynamical foundations from one proposition to another; it increased the number of basic methods for solving the direct problems. In place of the single method of the 1687 edition, Newton used in the radical revision three related but distinct methods of generating such solutions.

The first method, the linear dynamics ratio, is the initial measure of force that was introduced in Theorem 3 of *On Motion* and continued into Proposition 6 of the 1687 edition (i.e., $QR / QT^2 \times SP^2$). This measure of the force appears as Corollary 1 of the proposed Proposition 9, and it could have been applied to the same direct problems (circle/circumference, spiral/pole, and conics) as in the 1687 edition. The draft of Proposition 9 does close with the promise to provide examples of the procedure in the problems to follow, but they are not given. Whiteside speculates that if it had been Newton's intention to reproduce them, then he would have done so in a more abbreviated form than he used for the solutions in the 1687 edition.[10]

The second method, the circular dynamics ratio, was the technique alluded to in his early writings on curvature in the *Waste Book* in late 1664 or early 1665 when he wrote the following:

If the body b moved in an Ellipsis then its force in each point (if its motion in that point bee given) may bee found by a tangent circle of Equall crookednesse with that point of the Ellipsis.[11]

No examples of an application of curvature by Newton to the solution of elliptical motion have been found before the revisions of 1690. It is clear, however, that curvature played an important role in Newton's thoughts on dynamics, even here in 1664. It is tempting to speculate what use Newton did make of this suggestion before he developed the area law and the linear dynamics ratio in 1679. In 1690, the second measure of force, $1/SY^2 \times PV$, appears in Corollary 3 of the proposed Proposition 9 and Newton employed it to develop an entire set of alternate solutions for the exemplary problems that followed the initial Proposition 6 in the *Principia*. In contrast to the parabolic approximation of the first measure, in which a vanishingly small arc of an arbitrary curve is replaced by a vanishingly small parabolic arc, the second measure arose from a circular approximation, in which a vanishingly small arc of an arbitrary curve is replaced by a vanishingly small arc of the circle of curvature at that point. In figure 8.1, the line segment SY is the normal from the tangent to the center of force S and the line segment PV is the chord of the circle of curvature at the point P through the center of force S. The proposed Lemma 12, which followed this proposed Proposition 9, is concerned with chords of circles of curvature in conics, and stands in contrast to the initial Lemma 12, which is concerned with circumscribed areas about conics. In the proposed Proposition 10, the results of the previous proposition and lemma are applied to the problems that were set in the initial Propositions 7, 8, 10, 11, 12, and 13, and the solutions appear as Corollaries 1, 2, 3, and 4

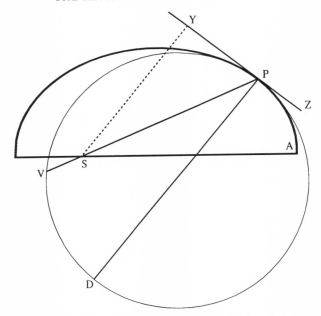

Figure 8.1 The circle DPV is the circle of curvature at point *P* of the general curve AP. It defines the chord *PV* and the diameter *PD*.

of the proposed Proposition 10, as Gregory noted when he wrote, "The proofs given in propositions 7 to 13 [of the 1687 *Principia*] inclusive now follow from it [the proposed Proposition 10] just like corollaries."

The third method, the comparison theorem, is a measure of force that gives the ratio of forces to two different force centers for any given orbit. The proposed Propositions 6 and 7 consider two different orbits with a common center of force and provide the basis for the demonstration in the corollary to the proposed Proposition 8. In that corollary Newton obtained the force directed to the center of an ellipse from that of the force directed to the center of a circle, which Gregory had reported as "this [force to the center of an ellipse is found] again from that of a constant centripetal force tending to the center of a circle." The proposed Proposition 8 extends the comparison from two given orbits with a single force center to a single given orbit with two different force centers. Thus, as Gregory reported of Newton's proposed revisions, one could obtain "the centripetal force of a body tending to the focus of a conic section from that of a centripetal force tending to the center." In Corollaries 2, 3, and 4 of the proposed Proposition 8, Newton gave solutions to the problems set for conics in the initial Propositions 11, 12, and 13.

NEWTON'S UNPUBLISHED PROPOSED REVISIONS   *173*

The first technique or method has been discussed in the analysis of Theorem 3 of *On Motion* in chapter 4 of this book and in the analysis of Proposition 6 in the 1687 edition in chapter 8. The second technique will be the subject of the analysis of the revised Proposition 6 in the 1713 and 1726 editions in chapter 9 to follow. The third technique, the new proposed Theorem 7, the comparison theorem, is the core of the proposed reconstruction, and it is carried forward to the revised published editions only in a disjointed form. To help the reader to identify it in its ultimate published but distorted form, it is presented below in the form Newton gave it in the proposed radical revision of the 1690s.

THE COMPARISON THEOREM

Figure 8.2A is based on the diagram for the new proposed Proposition 8, Theorem 7, in which Newton considers the ratio of the forces necessary to maintain motion in a single given orbit *ANPM* about two different centers of force, points *S* and *R*.[12] As with the linear dynamics ratio, the force is given as proportional to the ratio of the linear deviation and the square of the time. The two extracts from Newton's basic diagram in figure 8.2B show the deviations from the tangential motion as $Pf = yN$ and $Pe = xN$ (where the deviations yN and xN are parallel to the lines of force *PS* and *PR*) and the times are proportional to the shaded areas *SNM* and *RNM*. Thus, the ratio of the forces is given as follows:

$$F_S : F_R :: Pf / t_f^2 : Pe / t_e^2 :: Pf / SNM^2 : Pe / RNM^2$$

From similar triangles, Newton notes that the displacement *Pf* is proportional to the line *PS* and the displacement *Pe* is proportional to the line *PT*. Moreover, the area *SNM* is found to be proportional to PT^2 and the area *RNM* to PR^2:

$$F_S : F_R :: SP / PT^2 : PT / PR^2 :: SP \times PR^2 : PT^3$$

Thus, as the theorem states, the ratio of the forces is "as the product of the height of the first body *SP* and the square of the height of the second body *PR* . . . to the cube of the straight line *PT*" (i.e., $F_S / F_R = SP \times PR^2 / PT^3$).

The solution to the Kepler problem then follows in two short corollaries. In Corollary 1 the orbits are restricted to either circles or ellipses. Thus, if the force F_S is directed to the center of the ellipse, then it is known from the corollary to the new proposed Theorem 6 that it is proportional to *SP*. Thus, the force F_R directed to any other point is given by the following:

$$SP : F_R :: SP \times PR^2 : PT^3$$

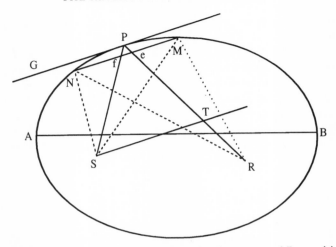

Figure 8.2A Based on Newton's diagram for the proposed Proposition 8.

Figure 8.2B The deviations *Pf* and *Pe* extracted from the diagram for the proposed Proposition 8.

or

$$F_R \propto PT^3 / PR^2$$

Thus, Newton obtained a general expression for the force F_R required to maintain elliptical motion directed to any general point R.

Now in Corollary 2 of the new proposed Proposition 8, Theorem 7, Newton obtains the solution to the distinguished Kepler problem simply by noting that if the point R is a focus of the ellipse, then the distance PT is a constant of the ellipse.[13] That result is all that is required to demonstrate that the force F_R is proportional to the inverse square of the distance from the body to the focal force center PR (i.e., $F \propto PT^3 / PR^2 \propto 1 / PR^2$, since PT is a constant). The solution is simplicity itself, particularly when contrasted to the solution in the 1687 edition that employs the linear dynamics ratio. Corollaries 3 and 4 of the new proposed Proposition 8 extend the result to parabolic and hyperbolic paths.

Thus, the new proposed Theorems 5 and 6 demonstrate that the force to the center of an ellipse is directly proportional to the distance, and the new Theorem 7 extends the result to a focal point and to all the other conic sections. The basics of the theorems flow smoothly out of the parabolic approximation and the area law. The proposed revision is a paradigm of directness and compactness.

THE PROPOSED PROPOSITIONS 9 AND 10

Newton was not satisfied in the projected revisions with simply giving the solution for the Kepler problem as obtained from the comparison ratio. In the first corollary of the new proposed Proposition 9 of the radical revision of 1694, he presents the original linear dynamics ratio of the initial Proposition 6 from the 1687 edition. Thus, he would have offered the solutions to the central and focal conic section direct problems that appeared in the 1687 edition as a second and complementary set of solutions to those provided by the new comparison theorem. More significant, however, is the appearance of a third set of solutions for the dynamic problems, the circular dynamics ratio, in the third and fifth corollary of the new proposed Proposition 9.

Following the new circular dynamics theorem of the proposed Proposition 9, Newton proposes a new Lemma 12. In all of the published editions Lemma 12 is concerned with the circumscribed area around an ellipse. The proposed version of Lemma 12, however, developed relationships between an ellipse and chords of its circle of curvature through the center and the focus of the ellipse. Both of these relationships are employed in the unpublished revisions to solve the series of problems concerned with circles and ellipses that appear in the first edition. In the published editions a separate proposition is used for each of the four problems concerning motion in a circle and in an ellipse that follow the basic dynamics theorem, Proposition 6. In the unpublished revisions, however, the solutions appear simply as four short corollaries to the new proposed Proposition 10. The charge of the proposed Proposition 10 is simply, "Let a body move in the perimeter of the conic PG: there is required the centripetal force tending to *any* given point S."[14] In obtaining a general answer, Newton employs a corollary from the proposed Lemma 12 and the circular dynamics ratio $1/YS^2 \times PV$ from the expanded version of the dynamics theorem, the proposed Proposition 9. These two results are combined quite simply to produce the same result that was found above as a corollary to the comparison theorem. Note here, however, that Newton does not need the solution of the central ellipse to obtain the solution for the focal ellipse. The solutions for both the central and focal ellipses come directly out of the new ratio. The force F_S, which is directed to any point S needed to maintain an ellipse, is given as follows:

$$F_S \propto PE^3 / PS^2$$

where the line PE is that portion of the chord of the circle of curvature cut by the transverse diameter DK, and PS is the line of force. The published solutions to Problems 2, 3, 5, and 6 appeared as simple three- or four-line corollaries to this proposed Proposition 10.[15]

Moreover, the proposed revisions contain a scholium in which Newton develops a general fluxional measure of the centripetal force. Whiteside claims, however, that although Newton quite skillfully developed the measure, he "twice failed accurately to apply it to the particular case of a focal conic."[16] Nevertheless, with the revised solution, Newton may have appeared to be almost too successful. The problems are solved so simply that one tends to forget the magnitude of the initial challenge.

THE LOCKE SOLUTION

In addition to the proposed radical revisions discussed above, an alternate approach to the direct Kepler problem exists in yet another manuscript, one that surfaced in the early 1690s. As noted earlier, the English philosopher John Locke expressed interest in the newly published 1687 edition of the *Principia* while he was in exile in Holland. It would appear, however, that his command of mathematics was limited because he sought assurance from the Dutch mathematician Christiaan Huygens that the geometry of the *Principia* was to be trusted.[17] Upon his return to England in 1689, Locke made Newton's acquaintance and "asked if the truth of the two fundamental propositions, namely, Propositions 1 and 11 in Book One, could not be demonstrated in some more simple way."[18]

Newton formally honored Locke's request for a "simpler solution" to these two propositions by sending him in 1690 a copy of a tract titled *A Demonstration that the Planets by their Gravity toward the Sun may move in Ellipses* (hereafter, *On Motion in Ellipses*).[19] It is difficult to see how the demonstration of Proposition 11 in this tract is simpler than the demonstration in the 1687 edition of the *Principia*. In fact, it appears that the tract is a copy of an earlier demonstration, and that Newton sent it to Locke rather than attempting to generate a simpler solution. If that surmise is correct, then there remains the question of the original source of the tract. Perhaps it was the version of the demonstration of the Kepler problem that Newton produced in 1679 prompted by his correspondence with Hooke on the topic of planetary motion (i.e., the elusive "lost solution of 1679") or perhaps it was the method alluded to in the early curvature statements of 1664.[20] When Halley visited Newton in Cambridge in August of 1684 and inquired after a solution for the problem of elliptical motion, Newton is reported to have replied that he had had such a solution but that he could not find it and so he would redo it. Newton's report of the discussion comes secondhand from the French mathematician Abraham Demoivre who, after Newton's death in 1727, told of a conversation that he had with Newton. Regarding Halley's request in August of 1684 for the 1679 solution, Demoivre's memorandum records the following:

> [Newton] looked among his papers but could not find it, but he promised him to renew it, & then to send it him. . . . In order to make good his promise he fell to work again, but he could not come to that conclusion wch he thought he had before examined with care. However he attempted a new way which tho^u longer than the first, brought him again to his former conclusion, then he examined carefully what might be the reason why the calculation he had undertaken before did not prove right, & he found that having drawn an Ellipsis coarsely with his own hand, he had drawn two Axes of the Curve, instead of . . . two Diameters somewhat inclined to one another, whereby he might have fixed his imagination to any two conjugate diameters, which was requisite he should do. That being perceived, he made both his calculations agree together.[21]

Thus, following the "lost solution" as remembered by Demoivre, there is a "first way" and a "new way." The solution done the first way presupposes conjugate diameters "somewhat inclined to one another." This first attempt to reproduce the lost solution fails because, as Newton states, the conjugate diameters are drawn incorrectly, inclined at right angles one to another. The solution done the new way appears not to introduce the conjugate diameters, because it is only after Newton has successfully completed the new solution that he is encouraged to return to the first calculation and *then* to discover the faulty construction of the conjugate diameters. The earliest existing solution of the problem of elliptical motion is found in Newton's tract *On Motion*, which Newton sent to Halley in 1684. It is quite different from the solution found in the tract *On Motion in Ellipses*, a version of which Newton sent to Locke in 1690. The solution that explicitly employs conjugate diameters appears in the 1687 edition of the *Principia*. The solution that does not explicitly employ conjugate diameters, the one done the "new way," may well be the source of the tract sent to Locke in 1690.

Turn now to the solutions for the problem of elliptical motion built upon the two different approaches. Figure 8.3 illustrates the solution to the problem that Newton sent to Halley in London in November of 1684 and that was eventually to appear in the first edition of the *Principia*. According to Demoivre, this solution appears to be Newton's first attempt because it uses the conjugate diameters *GP* and *DK*, which Newton first

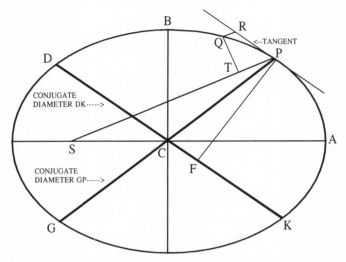

Figure 8.3 An ellipse *ABDGK* with conjugate diameters *PG* and *DK*.

drew incorrectly after Halley's visit. For any general point *P* on the ellipse, the diameter *DK* conjugate to *PCG* is constructed parallel to the tangent *PR* at the point *P*. In general, therefore, *GP* and *DK* will not be at right angles to each other. They must, as Newton noted, "be somewhat inclined to one another." The lines *BC* and *AC* are the two semi-axes of the curve and, in contrast to the conjugate diameters, must be at right angles. (It is the line *PF* that is at a right angle to *DK*, not *PG*.) Thus, the version sent to Halley shares with the first solution of 1684 and with the missing solution of 1679 an explicit dependence upon the conjugate diameters.

Consider now Newton's alternate solution of 1684, which employed the method "attempted in a new way" (i.e., the method that did not explicitly employ conjugate diameters). Only after "he attempted a new way which . . . brought him again to his former conclusion" did he "then examine carefully what might be the reason why the calculation he had undertaken before did not prove right." And it was then that he found his slip in drawing axes instead of conjugate diameters. Such an alternate solution, which is independent of the conjugate diameters, does not appear in the 1684 tract *On Motion* or in the 1687 edition of the *Principia*. It only appears in the solution given to Locke in 1690.

Consider first the special case given by Newton in Proposition 2 of the Locke solution.[22] Here Newton calculated the force of attraction at the two vertices of the ellipse's major axis. Inspection of the diagram in figure 8.4 will show that Newton makes no appeal to conjugate diameters. Note

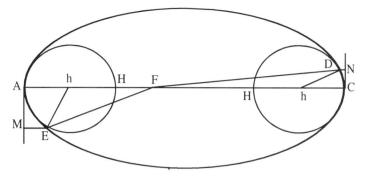

Figure 8.4 Based on Newton's diagram for Proposition 2 of the Locke solution.

that Newton's diagram contains only the axis, AC; there is no explicit rendering of the conjugate diameters. I have added the two circles of curvature at the perihelion, point A, and at the aphelion, point C. In the course of the solution, Newton is to call upon these two circles of curvature by specifying the following:

> because the Ellipsis is alike crooked at both ends those perpendiculars EM and DN will be to one another as the squares of the arches AE and CD.[23]

Now this statement is an echo of Newton's statement of 1664 found in his *Waste Book*.

> If the body moved in an Ellipsis then its force in each point . . . may be found by a tangent circle of equal crookednesse with that point of the Ellipsis.

The "alike crooked" and the "equal crookednesse" in the two statements refer to the curvature of the ellipse. The statements differ in that the first refers to the circle of curvature only at perihelion and aphelion while the second refers more generally to the circle of curvature at any point. One should not allow that difference, however, to obscure the possible connection. Whether or not Newton in 1690 had in mind his insight of twenty-five years before, there can be no doubt that, because conjugate diameters are nowhere alluded to in it, this part of the Locke solution could not have been the first solution and could well have been the alternative one generated by Newton in 1684. It is of interest to note, however, that the center of force F in this special case need not be the focus of the ellipse, the proof holds for any point on the major axis.[24]

In Proposition 3, which follows the special case of elliptical motion at aphelion and perihelion discussed above, Newton treats the general case of elliptical motion at any point under a central force directed toward the focus. As it was for the diagram for Proposition 2 (see fig. 8.4), Newton's

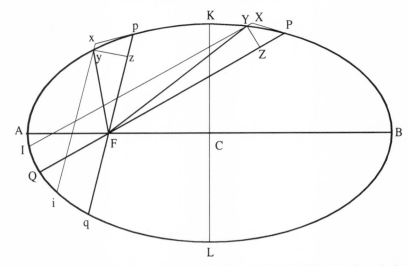

Figure 8.5 Based on Newton's diagram for Proposition 3 of the Locke solution.

diagram for Proposition 3 in figure 8.5 still does not contain an explicit rendering of the conjugate diameters. They do appear, however, in the diagram for Lemma 3 that precedes Proposition 3 and they are therefore employed in its solution. Thus, one can claim that Newton's demonstration of Proposition 3 is free of conjugate diameters only in the sense that he casts them out of the main frame of his argument of Proposition 3 and relegates them to his lemmatical subroutines. Thus, this solution lies somewhere between the solution in Proposition 2, which does not employ conjugate diameters in any fashion, and the solution in the first edition of the *Principia*, which employs them directly. If one accepts the testimony in Demoivre's memorandum, then Newton, following on Halley's visit to him in August 1684, could not find the solution of 1679 or, at first, even reproduce it. Instead he produced a variant solution that, unlike its predecessor, made no critical appeal to conjugate diameters. It is a solution that employs curvature as suggested in the statement of 1664, and may reflect Newton's thoughts on the problem long before his correspondence with Hooke in 1679.

CONCLUSION

The argument goes beyond the specific question of the nature of the lost solution of 1679. Far more interesting is the realization that Newton developed a number of interrelated but distinct methods of solving the Kepler problem; that these methods had their roots in Newton's earliest thoughts on dynamics; and that they grew to maturity, side by side. Unfortunately, the clear and uncluttered presentation of these methods set out in the proposed radical revisions of the early 1690s was never published. Instead, Newton elected to incorporate many of the additions and revisions into the existing structure of the 1687 edition of the *Principia*, and to publish the revised editions of 1713 and 1726 in that somewhat convoluted format. These revised published editions are the subject of chapter 9.

NINE

Newton's Published Recast Revisions
Two New Methods Concealed

Two more editions of the *Principia* were published during Newton's lifetime: the second edition appeared in 1713, twenty-six years after the first edition, and the third edition appeared in 1726, just one year before Newton died. Neither of these revised editions displays the radical restructuring of the propositions and lemmas that Newton had shown to the mathematician David Gregory in 1694. That proposed restructuring was set aside and Newton elected to retain the formal structure of the 1687 edition. All the propositions and lemmas of the revised editions retain their original headings with only minimal verbal changes. Many of the revised propositions, however, contain substantive additions to their text. In the revised editions, Newton tucked pieces of new theorems and new solutions into whatever existing nooks and crannies he could find in the formal structure of the 1687 edition. Thus, the jumbled published revision provides a much greater challenge to the reader than does the ordered unpublished revision discussed in chapter 8.

The key proposition for Newton's dynamics in all the published editions is Proposition 6, which sets out the basic paradigm for the solution of direct problems. In the 1687 edition, a single measure of the force was given in Proposition 6 by the linear dynamics ratio $QR / QT^2 \times SP^2$. In the proposed radical revisions of the 1690s, however, Newton offered two additional measures: the circular dynamics ratio, and the comparison theorem. The circular dynamics ratio $1 / SY^2 \times PV$ (where SY is the normal to the tangent through the center of force and PV is the chord of the circle of curvature through the center of force) appeared as Corollary 3 of Proposition 6 in the 1713 and 1726 editions and it was employed to provide alternate solutions for the direct problems in Propositions 7, 9, and 10. The alternate solution for the direct Kepler problem given in the revised Proposition 11, however, was obtained with the comparison theorem that Newton outlined in the unpublished revisions of the 1690s. In those proposed revisions, that technique was given the status of an independent theorem in a separate proposition. In the published revisions, however, it appears as a corollary at the conclusion of the direct problem for a circular orbit with a center of force at a general point (i.e., a revised version of Proposition 7). This revision permits Newton to retain the initial numbering sequence of the proposition, but it does so at a price: A simple example of how to employ the basic paradigm to solve direct problems now becomes a confusing combination of a solution to a general problem with a theorem hidden away in its final corollary.

The hallmark of the alternate measure of force is curvature. Newton's early work on curvature appears in his bound notebook, the *Waste Book*, in late 1664.[1] He was concerned with the problem of normals, curvature, and tangents. By May 1665 he had developed the general formula for the radius of curvature for a function $y = f(x)$.[2] By the winter of 1670–1671, Newton was writing a treatise, now called *Methods of Series and Fluxions*, that included as Problem 5 the "Determination of the Curvature of a Curve at any Point" and as Problem 6 the "Quality of Curvature at a Point." In this treatise he again derived the general measure of curvature in Cartesian coordinates.[3] He also provided a series of examples in which he calculated the radius of curvature of various curves, including that of the three conic sections.[4] He concluded, "Since all geometrical curves . . . can be referred to right-angled ordinates, I believe I have done enough. Anyone who desires more will provide it without difficulty by his own efforts, especially if, as a bonus, in illustration of the point I add a method for spirals."[5] Then he developed an expression for the radius of curvature in polar coordinates, and he again offered a series of examples in which he demonstrated how the technique could be applied to specific curves.[6]

Newton did not employ any of the mathematical details of curvature in his revised dynamics. The concept of curvature, however, provided the very basis for the alternate measure of force. Just as the key to Newton's dynamics in the first edition is uniform *rectilinear* accelerated motion, so the key to his dynamics in the revised editions is uniform *circular* motion. In both of the revised published editions, the second edition of 1713 and the third edition of 1726, Newton retained the linear dynamics ratio $QR/QT^2 \times SP^2$ (based on the parabolic approximation) as his fundamental device for solving the direct problem. But in the revised editions he extended Proposition 6 by the addition of a second method of analysis: a circular dynamics ratio, which is based on uniform circular motion. It stands in contrast to the linear dynamics ratio, which was based upon uniform linear motion.

Of the many revisions and additions that appear in the later editions of

184 THE REVISIONS AND EXTENSIONS

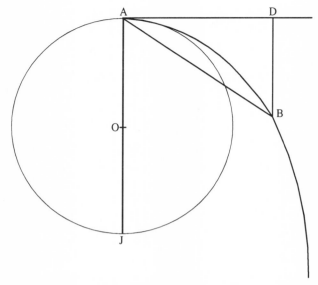

Figure 9.1 Lemma 11: the circle *AJ* represents the circle of curvature at point *A* of the general curve *AB*.

the *Principia*, three are of particular importance in discussing Newton's dynamics: first, the revision of Lemma 11 by the insertion of a curvature qualification and the addition of two new corollaries; second, the revision of Proposition 1 by the addition of six corollaries (a defense of the polygonal approximation); and third, the revision of Proposition 6 by changing its basic time dependence from Lemma 10 (based on rectilinear motion) to Lemma 11 (based on circular motion) and the addition of the circular dynamics ratio.

THE REVISION OF LEMMA 11

Figure 9.1 is based upon Newton's diagram for Lemma 11, where the circle *AJ* represents the circle of curvature at point *A* on the general curve *AB*. The specific problem faced in Lemma 11 is the demonstration that the chord *AB* of the general curve *AB* is proportional to the square of the line segment *BD* and the assertion that the ratio BD^2 / AB remains finite as point *B* approaches point *A*. If the motion of a body along an infinitesimal arc at point *A* on the general curve *BA* is represented by motion along an infinitesimal arc of its circle of curvature *AJ* at that point, then the assertion that ratio BD^2 / AB remains finite means that the general curve has a

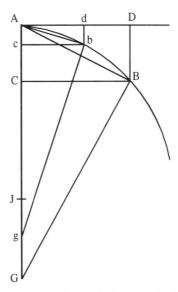

Figure 9.2 Based on Newton's diagram for Lemma 11.

finite non-zero curvature at the point A (i.e., that the diameter AJ of the circle of curvature at the given point A is finite).

Figure 9.2 is based on Newton's diagram for Lemma 11 and it remained unchanged through the three editions of the *Principia*. In the 1687 edition Newton employed Lemma 11 in Propositions 4 and 9. The former derives the nature of the force directed to the center of the circle required to maintain uniform circular motion (force $\propto$ speed^2 / SP), and the latter derives the nature of the force in the direct spiral/pole problem (force $\propto$ 1 / SP^3).[7] In the revised editions, however, he no longer employed Lemma 11 in Proposition 4. Instead, he used the new corollaries that had been added to Proposition 1. He promoted Lemma 11, however, to the important role of establishing the dependence of the line segment QR upon the square of the time in the derivation of the original dynamics ratio in Proposition 6. Lemma 10, which served that function in the 1687 edition, was given a secondary role in the revised editions.

The statement of Lemma 11 as it appears in the revised editions is identical to the statement of the lemma that appears in the first edition, except that Newton added a qualifying phrase concerning the finite curvature.

[Lemma 11] *The vanishing subtense of the angle of contact [the line BD], in all curves, which at the point of contact have a finite curvature, is ultimately in the doubled ratio of the subtense of the conterminate arc [the arc AB].*

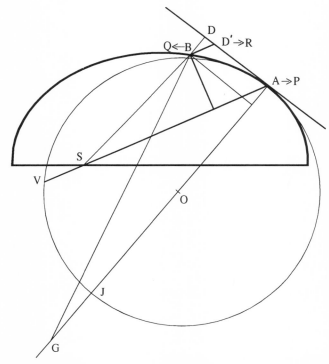

Figure 9.3 The diagram for Lemma 11 adapted to the diagram for Proposition 6.

More significant than the explicit reference to curvature, however, is Newton's addition of a qualification into Case 3 and the addition of two new corollaries. In both the first and the revised editions, Case 2 demonstrates that the ultimate ratio of the line BD to the square of the arc AB^2 will be the same even if the line BD "be inclined to AD in any given angle." This result is important in Proposition 6 because the line segment DB is to become the inclined linear deviation QR that appears in the dynamics ratio, $QR / QT^2 \times SP^2$. In addition to being inclined with respect to the tangent, however, the displacement QR changes along the curve as a function of the equation that generates the curve. To accommodate that behavior, Newton revises Lemma 11 by inserting in Case 3 the additional qualification that DB (and hence QR) can be "determined by any other condition whatever."

Figure 9.3 is a revision of the diagram for Lemma 11 into the form that applies to Proposition 6. Thus, the inclined line BD' is to become the displacement QR and the point A to become the point P. The new Corollary 2 states that the line AC will become as the line BD and hence, from the opening proof, as the square of the arc *AB*. Now the new Corollary 3 will relate the arc *AB* to the time and hence will relate the line *AC*, and ultimately the inclined line *BD'*, to the square of the time. This dependence upon time, however, is inserted by Newton with very little warning.

> Corollary 3. *And therefore the versed sine*[8] [*the line* AC] *is in the doubled ratio of the time in which a body will describe the arc with a given velocity.*

The "therefore" may require some reflection on the part of the reader. As Whiteside puts it, "The tacit assumption here made is that the small arcs *AB* and *Ab* may adequately be approximated by corresponding arcs of their osculating circles."[9] But this corollary goes beyond simply replacing the arc of a general curve by the arc of the circle of curvature; it replaces the general motion over the arc of the curve with *uniform circular motion* over the arc of the circle of curvature. The displacement *BD'* is proportional to the square of the arc and if the "body will describe the arc with a given velocity" (i.e., if the body moves with uniform circular motion), then the arc is proportional to the time *t*, and hence the displacement *BD'* is proportional to the square of the time. It is this corollary of Lemma 11 that Newton employs in the revised editions to establish *both* the linear and circular dynamics ratios.

THE REVISION OF PROPOSITION 1

The statement of Proposition 1 in the revised editions contains the added qualification that the center of force and the plane of the motion are fixed. Otherwise, the demonstration of the basic proposition for equal areas in equal times remains the same as in the 1687 edition. Newton dramatically revised, however, the number and content of the corollaries that follow the basic proposition. The 1687 edition contains only two corollaries and they set out the conditions under which the areas were not proportional to the times (i.e., motion under noncentral forces). In the revised editions, Newton replaces these two corollaries with six new and different corollaries.

The first three of the new corollaries at last provide a defense of Newton's basic use of the parallelogram rule (*On Motion*: Hypothesis 3; *Principia*: Corollary 1) in obtaining the fundamental measure for an impulsive force (i.e., the observed displacement is the diagonal of the parallelogram formed by the tangential displacement and the radial deviation).[10] Newton employs this technique in the polygonal approximation in his earliest analysis of circular motion, and he continued to employ it in the tract *On Motion* and in the 1687 *Principia*. Newton had not previously given a formal defense of it, although I have discussed it in considerable detail in

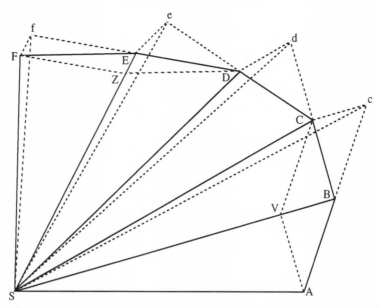

Figure 9.4 Based on Newton's diagram for Proposition 1 in the revised editions.

chapters 1, 2, and 4. The following are the new corollaries added to Proposition 1 in the revised editions of the *Principia*, and figure 9.4 is based on the figure for Proposition 1 that appears in the revised editions.

> Corollary 1. *In nonresisting spaces, the velocity of a body attracted to an immobile center is inversely as the perpendicular dropped from that center to the straight line which is tangent to the orbit. For the velocities in those places* A, B, C, D, *and* E *are respectively as the bases of the equal triangles* AB, BC, CD, DE, *and* EF, *and these bases are inversely as the perpendiculars dropped to them.*

The equal areas of the triangles are given by the product of the base of the triangle, which is the velocity multiplied by the equal times, and the slant height of the triangle, which is the perpendicular to the tangent through the center of force. Thus, the product of the velocity and the tangent is equal to the given area divided by the given time and, thus, the velocity is inversely proportional to the perpendicular to the tangent Y (i.e., area ∝ base x height = velocity x time x Y, or velocity v ∝ area/time x Y, where the area/time is a constant of the motion).

> Corollary 2. *If chords* AB *and* BC *of two arcs successively described by the same body in equal times in nonresisting spaces are completed into the parallelogram* ABCV, *and diagonal* BV (*in the position that it ultimately has when those arcs are decreased indefinitely*) *is produced in both directions, it will pass through the center of forces.*

The line *Bc* is constructed equal in length to the line *AB* and the line *Cc* is constructed parallel to the line of force *BS*. Thus, the diagonal of the parallelogram *ABCV* will pass through the center of force *S*.

> Corollary 3. *If chords* AB, BC *and* DE, EF *of arcs described in equal times in nonresisting spaces are completed into parallelograms* ABCV *and* DEFZ, *then the forces at* B *and* E *are to each other in the ultimate ratio of the diagonals* BV *and* EZ *when the arcs are decreased indefinitely. For the motions* BC *and* EF *of the body are (by Corollary 1 of the laws) compounded of the motions* Bc, BV *and* Ef, EZ; *but in the proof of this proposition* BV *and* EZ, *equal to* Cc *and* Ff, *were generated by the impulses of the centripetal force at* B *and* E, *and thus are proportional to these impulses.*

If there were no impulsive force at point *B*, then the body would make the displacement *Bc* by virtue of its velocity at *B*. If the body had no initial velocity at *B*, then it would make the displacement *BV* by virtue of the impulsive force at *B*. Thus, by "Corollary 1 of the laws" (i.e., the parallelogram rule) the displacement *BC* is "compounded of the motions *Bc* and *BV*." The velocity increment Δv at *B* generated by the impulsive force is the product of the force F_B and the extremely short time of the impulse δt. The displacement *BV* generated by that velocity increment is the product of the velocity increment Δv and the time between impulses ΔT. Thus, $BV = (\Delta_v)(\Delta T) = F_B(\delta t)(\Delta T)$ and the displacement $EZ = F_E(\delta t)(\Delta T)$. The ratio of the displacements at points *B* and *E* is given by $BV/EZ = F_B(\delta t)(\Delta T) / F_E(\delta t)(\Delta T) = F_B / F_E$, or "the forces at *B* and *E* are to each other in the ultimate ratio of the diagonals *BV* and *EZ*." This ratio of displacements as a measure of the ratio of forces has been fundamental to all of Newton's dynamics. In fact, it has been so fundamental that until this late date he appeared to think it was not necessary to defend it.

> Corollary 4. *The forces by which any bodies in nonresisting spaces are drawn back from rectilinear motions and are deflected into curved orbits are to one another as those sagittas of arcs described in equal times which converge to the center of forces and bisect the chords when the arcs are decreased indefinitely. For these sagittas are halves of the diagonals with which we dealt in Corollary 3.*

Figure 9.5 is fashioned after the drawing in Whiteside's discussion of these corollaries.[11] The "sagitta of arc" is the line segment *Bx*, where the sagitta is "the arrow in the bow" of the orbital arc *ABC*. In terms of the trigonometric functions of a unit circle, the sagitta is the versine (i.e., 1 − cosine).[12] The preceding corollary relates to impulsive forces, as employed in the polygonal approximation. This corollary relates to continuous forces (i.e., "when the arcs are decreased indefinitely"), as employed in the parabolic approximation. The revised Proposition 6, which demonstrates both the linear and circular dynamics ratio, makes specific reference to this new Corollary 4 of Proposition 1 in relating the force to the displacement (in contrast to the 1687 edition, which makes reference to the second law,

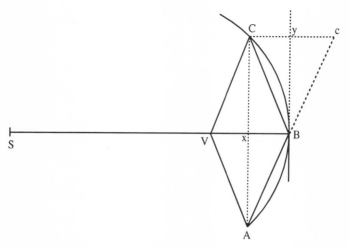

Figure 9.5 The sagitta of arc is the line segment *Bx* (the "arrow") in the arc *ABC* (the "bow").

F = ma). The sagitta *Bx* is equal to the segment *Cy* and will become the deviation *QR* in the linear dynamics ratio.

> Corollary 5. *And therefore these forces are to the force of gravity as these sagittas are to the sagittas, perpendicular to the horizon, of the parabolic arcs that projectiles describe in the same time.*

The body at point *B* has an initial velocity in the direction *By* and the impulsive force acts perpendicularly to that velocity and is directed toward the center of force *S*. In the limit of small arcs, the force is constant, as is the force of gravity near the surface of the earth. As projectiles fired horizontally move in parabolic arcs under the constant vertical force of gravity, so the arcs of these sagittas are parabolic under the constant force toward *S*. Here Newton makes explicit the parabolic approximation discussed here in chapter 2 and which he employed implicitly in much of *On Motion* and the first edition of the *Principia*.

> Corollary 6. *All the same things hold, by Corollary 5 of the laws, when the planes in which the bodies are moving, together with the centers of forces that are situated in those planes, are not at rest but move uniformly straight forward.*

THE REVISION OF PROPOSITION 6

In the first edition Newton set out the nature of the deviation *QR* quite simply: "The nascent line segment *QR* is, given the time, as the centripetal force (by Law 2), and, given the force, as the square of the time (by Lemma 10)." In the revised editions, however, Law 2 is replaced by a reference to the new corollary from Proposition 1, and Lemma 10 is replaced by a reference to Lemma 11.

First Edition	Revised Editions
In the indefinitely small figure QRPT *the nascent line segment* QR *is, given the time, as the centripetal force (by Law 2),*	*For the sagitta* [QR] *in a given time is as the force (by Proposition 1, Corollary 4), and on increasing the time in any ratio, because the arc is increased in the same ratio, the sagitta* [QR]
and, given the force, as the square of the time (by Lemma 10);	*is increased in that ratio doubled (by Corollaries 2 and 3 of Lemma 11),*
and hence, when neither is given, as the centripetal force and the square of the time conjointly;	*and hence, it is as the force and the square of the time jointly. Let the doubled ratio of the time be taken away from each side,*
and hence the centripetal force is as the line segment QR *directly, and the square of the time inversely.*	*and the force will be as the sagitta* [QR] *directly and the square of the time inversely.*

Thus, the time squared dependence of the displacement QR (now called the "sagitta" or the "versine") stems directly from Corollary 3 of Lemma 11 (i.e., the circular approximation discussed above). When the proof of this statement was finished, Newton added the following statement: "The same thing is easily demonstrated by Corollary 4 of Lemma 10," which was the only reference given in the 1687 edition. He has now relegated the linear demonstration of Lemma 10 to a secondary position relative to the circular demonstration of Lemma 11.

Newton then continued to develop the linear dynamics ratio $QR/QT^2 \times SP^2$ as in the first edition. In the revised editions, this ratio continues to serve as the primary method of solving the direct problems that follow Proposition 6, including the direct Kepler problem, as it did in the 1687 edition. It is significant, however, that in the revised version of Proposition 6 even the linear dynamics ratio rests upon the approximation to uniform circular motion given in Lemma 11.

But even more dramatic is the introduction of an entirely new and alternate dynamics ratio. Figure 9.6 is a comparison of the diagram from the 1687 edition with the new diagram that accompanies Proposition 6 in the revised editions. The most obvious change is the addition of the line YS, which passes through the force center S and is normal to the tangent

192 THE REVISIONS AND EXTENSIONS

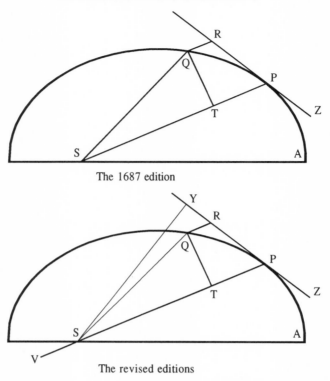

Figure 9.6 A comparison of Newton's diagrams for Proposition 6. In the revised edition the line *PS* is extended to the point *V* and the perpendicular to the tangent *YS* is added.

YZ. Newton has demonstrated in the first of the new corollaries added to Proposition 1 that the line *YS* is inversely proportional to the speed of the body at point *P*. A more subtle but significant change in the figure is the extension of the line of force *SP* through the force center *S* to a point *V*. This extended line *PV* is identified in the text as the chord of the circle of curvature at the point *P* that passes through the center of force *S*. When that circle is added to the diagram, as in figure 9.7, then the nature of the extension of the line of force to the point *V* is made clear. The new Corollary 3 to Proposition 6 relates the centripetal force to the chord of the circle of curvature *PV* and the perpendicular to the tangent *SY* as follows:

> Corollary 3. *If* PV *be a chord of this circle, drawn from the body through the center of force;* [then] *the centripetal force will be reciprocally as the solid* SY² x PV.

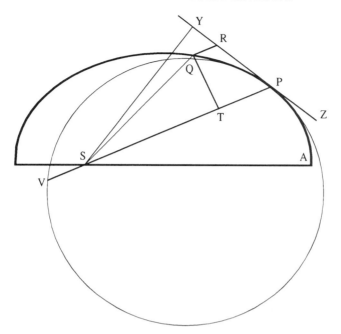

Figure 9.7 A revision of Newton's diagram for Proposition 6 with the circle of curvature *PQV* added.

Thus, in the revised editions Newton introduced a second measure of the force, the circular dynamics ratio $1 / SY^2 \times PV$, to complement the linear dynamics ratio, which is the only measure in the 1687 edition. The significant distinction between the linear and the circular version of the dynamics ratio is that the former was developed in part from considerations of uniform rectilinear motion, while the latter was developed from considerations of uniform circular motion. In the published version, the revision appears as a logical extension of the original. In the unpublished radical restructurings, however, the revision took on a much more significant role. There are indications, in fact, that Newton toyed with the idea of making the circular dynamics ratio the primary measure of force (i.e., reversing the order of presentation).[13] In the following, I give the entire revised Proposition 6 of the 1713 and 1726 editions. The details of the demonstration of the body of the proposition and the derivation of the linear dynamics ratio in Corollary 1 are similar to the analysis in the 1687 edition and in Theorem 3 of the tract *On Motion*, which I discussed in detail in chapter 4. I here restrict my detailed comments to the revised corollaries.

PROPOSITION 6. THEOREM 5

Proposition 6. *If a body in a nonresisting space should revolve in any orbit around an immobile center and should describe any just nascent arc in a minimal time, and if the sagitta of the arc should be understood to be drawn so that it would bisect the chord, and when produced, would pass through the center of the forces;* [then] *the force in the middle of the arc will be as the sagitta directly and the square of the time inversely.*

For the sagitta in a given time is as the force (*by Proposition 1, Corollary 4*), and on increasing the time in any ratio, because the arc is increased in the same ratio, the sagitta is increased in that ratio doubled (*by Corollaries 2 and 3 of Lemma 11*), and thus it is as the force and the square of the time jointly. Let the doubled ratio of the time be taken away from each side, and the force will be as the sagitta directly and the square of the time inversely. Which was to be proven.

The same thing is easily demonstrated by *Corollary 4 of Lemma 10*.

Corollary 1. *If a body P by revolving around the center S, should describe any curved line APQ, and if the straight line ZPR should touch that curve at any point P, and if to this tangent from any other point Q of the curve, QR should be drawn parallel to the distance SP, and if QT should be dropped perpendicular to the distance SP;* [then] *I assert that the centripetal force will be reciprocally as the solid* SP^2 × QT^2 / QR, *provided that the quantity of that solid that ultimately occurs when the points P and Q coalesce is always taken.* For QR is equal to the sagitta of double the arc QP, in the middle of which is P, and the double of the triangle SQP, or SP × QT, is proportional to the time in which that double arc is described; and thus it can be written as an expression of the time.

Corollary 2. *By the same argument the centripetal force is reciprocally as the solid* SY^2 × QP^2 / QR, *if only there is constructed the perpendicular SY dropped from the center of forces onto the tangent PR of the orbit. For the rectangles SY × QP and SP × QT are equal.*

See figure 9.8. As the point Q approaches the point P, then the arc QP approaches the tangent segment RP. Thus, the triangles TQP and YSP become similar and therefore $SY / SP = QT / QP$ or as Newton states it, "the rectangles $SY \times QP$ and $SP \times QT$ are equal." The linear dynamics ratio $QR / (QT^2 \times SP^2)$ thus can be written as $QR / (SY^2 \times QP^2)$.

Corollary 3. *If the orbit is either a circle, or touches a circle concentrically, or cuts it concentrically, that is, contains the minimal angle of contact or of section, having the same curvature and the same radius of curvature at point P; and if there is constructed the chord PV of this circle, drawn from the body through the center of forces;* [then] *the centripetal force will be reciprocally as the solid* SY^2 × PV; *for* PV *is* QP^2 / QR.

Figure 9.9A is a diagram based on Euclid, Book 3, Proposition 35, in which it is demonstrated that the products $AE \times EC$ and $BE \times ED$ of the segments of the chords of a circle are equal (or from Proposition 15, Book 1 of Apollonius's *Conics*; see figure 5.12). Figure 9.9B is a diagram of that Euclidian proposition applied to the circle of curvature at a point P of a general

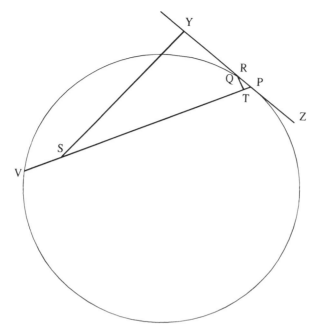

Figure 9.8 As the point Q approaches the point P the angle TQP approaches the angle YSP, and triangles TQP and YSP are similar.

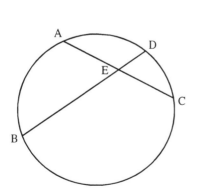

Figure 9.9A Based on Proposition 35, Book 3 of Euclid's *Elements*: $AE \times EC = BE \times ED$.

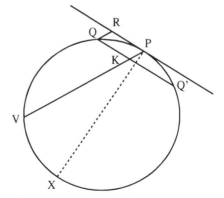

Figure 9.9B Euclid's Proposition 35 applied to the circle of curvature in Newton's Proposition 6: $QK \times KQ' = VK \times KP$.

curve (not shown). Thus, $QK \times KQ' = PK \times KP$ where $KP = QR$ because $PRQK$ is a parallelogram by construction. In the limit as the point Q approaches the point P, then PK approaches PV and QK approaches KQ' or QP. Thus, $QP \times QP = PV \times QR$ or, as Newton states, "PV is as QP^2 / QR." Thus, the reciprocal measure of the force from Corollary 2, $(SY^2 \times QP^2)/QR$, can be written as $SY^2 \times (QP^2 / QR)$ or $SY^2 \times PV$.

> Corollary 4. *With the same suppositions, the centripetal force is as the square of the velocity directly and the chord inversely. For the velocity is reciprocally as the perpendicular* SY, *by Corollary 1 of Proposition 1.*

From Corollary 3, the force is proportional to $1/SY^2 \times PV$ and from Corollary 1 of Proposition 1, the velocity v is inversely proportional to SY, the perpendicular to the tangent through the center of force S. Thus, the force is proportional to v^2 / PV.

> Corollary 5. *Hence if any curvilinear figure* APQ *is given, and on it a point* S *is also given, toward which a centripetal force is perpetually directed, there can be found a law of centripetal force, whereby any body* P, *perpetually drawn back from a rectilinear course, will be confined in the perimeter of that figure and will describe it by its revolution. Of course there must be computed either the solid* $SP^2 \times QT^2$ *over* QR *or the solid* $SY^2 \times PV$ *reciprocally proportional to this force. We shall give examples of this in the following problems.*

In this demonstration of Proposition 6, the circular dynamics ratio is derived from the linear dynamics ratio. Newton has elsewhere reversed the procedure and obtained the circular dynamics ratio directly without any reference to the linear ratio.

A REJECTED REVISION OF PROPOSITION 6

Newton produced a number of variations of Proposition 6 that never appeared in print. In one draft, he crossed out the final corollaries and inserted the title "Prop VI." He then proceeded to derive the circular dynamics ration from first principles, without any reference to the linear dynamics ratio.[14] Figure 9.10 is a reconstruction of the small sketch that appears in that revised draft version of Proposition 6. The symbols are as in the published edition: P for the place of the body, R for a point beyond P on the orbit, and v and V for the terminal of the chords drawn from P through S of the two circles that touch the orbit at P. After much revision and deletion, Newton's draft of this version of Proposition 6 is as follows.

> Proposition 6. *If circles touch orbits concave to the bodies and if they are of the same curvature with the orbits on the points of contacts,* [then] *the forces will be reciprocally as the solids comprised of the chords of the arcs of circles from bodies through the centers of forces and by the squares of the perpendiculars descending from the same centers on to the rectilinear tangents.*

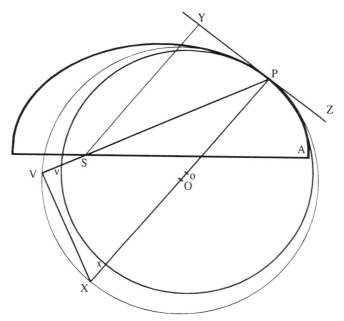

Figure 9.10 A reconstruction from Newton's sketch for the proposed revision of Corollary 6 of Proposition 6.

> *Let S be the center of forces, the body P revolving on the orbit PR, the circle[s] PVX [and Pvx] touch the orbit at P, at the concave parts of it; whichever [circle] is of the same curvature with the orbit at the point of contact, the chord PV of this circle constructed from the body P through the center S; the straight line PY touching the orbit at P and SY perpendicular from the center S falling on this tangent: I say that the centripetal force of the revolving body is reciprocally as the solid* $SY^2 \times PV$.[15]

Newton does not give details of the demonstration, but it is not difficult to reconstruct a version of one. The basic assumption is that the force F_O (directed toward the center O) is in proportion to the force F_S (directed toward the center S) as the chord of the circle of curvature PV drawn through the center S is in proportion to the diameter of the circle of curvature PX (i.e., $F_O / F_S = PV / PX$).[16] From Proposition 4, the force F_O required for uniform circular motion is directly proportional to the square of the speed (or inversely as the perpendicular SY) at point P and inversely proportional to the radius (or the diameter PX) of curvature, or $F_O \propto$ (speed)^2 / (radius). Thus, one has that $F_S = [F_O](PX/PV) \propto [1 / (SY^2 \times PX)] (PX / PV) = [1/(SY^2 \times PV)]$, and thus, as Newton puts it "the centripetal force of the revolving body is reciprocally as the solid $SY^2 \times PV$."[17]

Newton did not adopt such a radical revision, however, and in the published editions the circular dynamics ratio is derived from the linear dynamics ratio rather than from first principles. The circular dynamics ratio appears as an alternate measure of force rather than as the primary measure. Thus, Newton names the solutions to the direct problems that employ it as "alternate solutions."

In the 1687 edition Newton fulfilled his promise to "give examples of this [the linear dynamics ratio] in the following problems" by considering three preliminary examples (circle/circumference, spiral/pole, and ellipse/center) before presenting the solution to the direct problem of motion in a conic with the center of force at a focus. In the revised editions he considers the same three preliminary examples as in the 1687 edition but now the antecedent of "this" in his statement refers to both the linear dynamics ratio and the circular dynamics ratio. The solutions employing the former are essentially unchanged from the versions that appear in the 1687 edition, and the solutions employing the latter follow each example as "alternate solutions."

THE REVISION OF PROPOSITION 7: AN ALTERNATE SOLUTION

The problem that immediately followed the general paradigm for solving direct problems in *On Motion* and in the first edition of the *Principia* was the direct problem of a circular orbit with the center of force on the circumference of the circle. The problem had no physical application, but it was a very simple and instructive example of how the paradigm was to be applied (see Problem 1 in chapter 5). In the revised edition, however, the nature of the problem changes: the force center is a general point and the solution to the original problem appears in a corollary. The statement of Proposition 7 in the first versus the revised editions is as follows:

First Edition	Revised Editions
Let a body be orbited on the circumference of a circle; there is required the law of the centripetal force being directed to some given point on the circumference.	Let a body be orbited on the circumference of a circle; there is required the law of the centripetal force being directed to any given point.

The statement in the revised editions is a major departure from the statement in the 1687 edition; the force center is no longer restricted to a point *on the circumference* of the circle but may be located at *any point*. Moreover, in the first edition, Proposition 7 consisted of about a dozen lines in total. In the revised editions it has increased in length some fivefold. The en-

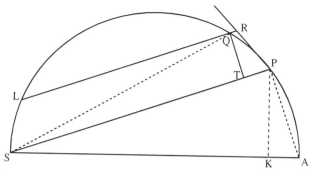

Figure 9.11A Based on Newton's diagram for Proposition 7 in the 1687 *Principia*.

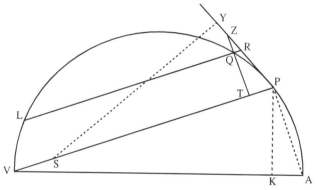

Figure 9.11B Based on Newton's diagram for Proposition 7 in the revised *Principia*.

larged proposition still presents a solution using the linear dynamics ratio, but Newton has added an alternate solution using the circular dynamics ratio. Newton also has added three corollaries: the first corollary gives the solution to the circle/circumference problem that appears as Corollary 1; the second corollary gives the ratio of forces directed toward two different centers of force for the same given circular orbit; and the third corollary extends that result for a given *circular* orbit to *any* given orbit.

Figure 9.11A is based on Newton's diagram that accompanies Proposition 7 in the 1687 edition, and Figure 9.11B is based on his diagram for Proposition 7 in the revised editions. Again, the most obvious change is the addition of the line *YS*, the normal to the tangent that is employed in the circular dynamics ratio. One notes also that the center of force *S* has shifted from a point restricted to the circumference of the circle to some general point, as is dictated by the revised statement of the problem. Finally, the line of force from point P has been extended through the force center S to the circle at a point V. In this very special problem of a circular orbit, the orbit is identical with its circle of curvature. Thus, PV is the chord of the circle of curvature, as required in the circular dynamics ratio. The radius of the circle, OP, should be added to Newton's figure to aid in the analysis where O is the midpoint of VA.

Just as the procedure for solving the direct problems with the linear dynamics ratio can be reduced to a series of three simple steps, so can the procedure using the circular dynamics ratio be reduced to the following simple paradigm:

Step 1 Seek out an expression for the perpendicular to the tangent SY,
Step 2 Seek out an expression for the chord of the circle of curvature PV, and
Step 3 Combine them in the revised dynamics ratio $1 / SY^2 \times PV$ to obtain the force.

Note that one need no longer submit the circular dynamics ratio to a limiting procedure as was necessary with the linear dynamics ratio. In passing from the parabolic approximation to the circular approximation, Newton has already invoked an effective limiting operation. The force is given directly by the chord PV and the square of the perpendicular YS as they exist in the diagram. The alternate solution for Problem 2 requires only a few lines to complete the paradigm.

Step 1. Find "SY." From the similarity of triangles SYP and VPA in figure 9.11, Newton obtains the following relationship:[18]

$$SY = SP \times PV / AV$$

Step 2. Find "PV." The expression for the chord PV already exists in the expression for the normal SY in Step One.

Step 3. Find "the ratio." Thus, if one squares both sides of the expression above and multiplies by PV, then the circular dynamics ratio provides the following expression:

$$F \propto 1 / SY^2 \times PV = AV^2 / SP^2 \times PV^3 \propto 1 / SP^2 \times PV^3$$

where AV is the diameter of the circle and hence a given constant. But the solution for the new problem (i.e., $F \propto 1 / SP^2 \times PV^3$) is not expressed in terms of the line of force SP alone, as are all of the other solutions of direct problems. The different type of result for the revised Proposition 7 occurs because the force is directed to *any* point, not simply to a *specific* point, as in the earlier version of the problem (specifically, a point on the circumference of the circle).

Newton extends the general nature of the solution for the circular orbit in Proposition 7 farther by adding new corollaries. In Corollary 1 he notes that as the general point S moves to the circumference of the circle, the chord PV is equal to the line of force SP, and hence the result is that the force goes directly as the fifth power of SP, as he demonstrated in the version of the problem that appeared in the first edition. In Corollary 2 Newton extends the scope of the problem by calculating the ratio of the forces directed toward any *two* arbitrary points that will maintain the circular orbit.

> Corollary 2. *The force by which the body P in the circle* APTV *revolves about the center of force* S *is to the force by which the same body P may revolve in the same circle and the same periodic time about any other center of force* R . . .

Corollary 3 makes the final extension: the *circular* orbit becomes *any* orbit.

> Corollary 3. *The force by which the body P in any orbit revolves about the center of force* S, *is to the force by which the same body may revolve in the same orbit, and the same periodic time about any other center of force* R, *as the solid* SP × RP².

The final line of the corollary provides Newton's *only* formal defense for extending the solution for a problem in circular motion into a dynamics theorem for general motion.

> [Corollary 3] *For the force in this orbit at any point* P *is the same as in a circle of the same curvature.*

Thus, the conclusion of Proposition 7 provides yet another alternative source of solutions for dynamic problems, the comparison dynamics ratio. In the proposed revision of the 1690s it appeared prominently as Theorem 7 in the proposed Proposition 8. In the published revisions, it is tucked away as a corollary to Problem 2 in Proposition 7. The comparison dynamics ratio is a result that rests upon the circular approximation of Lemma 11 but it is a variation of the revised version that appears in Theorem 5. Corollary 3 itself is worthy of the title "theorem," but it appears simply as an extension of a problem. Note that Newton cannot use the comparison dynamics ratio of Corollary 3, Proposition 7, to solve the problems that directly follow it in Section 2 (i.e., the spiral/pole of Proposition 9 and the ellipse/center of Proposition 10) because he does not have a comparison solution for them as he did for Proposition 11. In Propositions 9 and 10 he employs the circular dynamics ratio from Proposition 6 to produce the alternate solution. Newton employs the unheralded comparison dynamics ratio, however, in the alternate solution of the direct Kepler problem in Proposition 11 when he compares the solution of the ellipse/center (Proposition 10) to the solution of the ellipse/focus (Proposition 11).

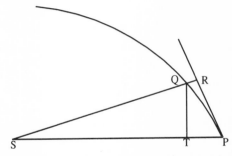

Figure 9.12A Based on Newton's diagram for Proposition 9 in the 1687 *Principia*.

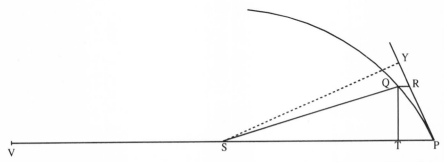

Figure 9.12B Based on Newton's diagram for Proposition 9 in the revised *Principia*.

THE REVISION OF PROPOSITION 9: AN ALTERNATE SOLUTION

In this direct problem, the orbit is an equal angle spiral and the center of force is located at the pole of the spiral. The statement of the problem and the solution employing the linear dynamics ratio remain as they were in the 1687 edition. That initial solution is brief compared to the solutions of the other problems that followed it. (See chapter 8 for a full discussion of the initial solution.) The alternate solution employing the circular dynamics ratio, however, is even more brief, requiring only the following few lines.

> Proposition 9. *The perpendicular* SY *dropped to the tangent, and the chord* PV *of the circle cutting the spiral concentrically are to the distance* SP *in given ratios; and thus* SP³ *is as* SY² x PV, *that is* (*by Proposition 6, Corollaries 3 and 5*), *reciprocally as the centripetal force.*

Figures 9.12A and 9.12B compare Newton's diagrams for Proposition 9 in the 1687 edition and the revised editions. The normal to the tangent

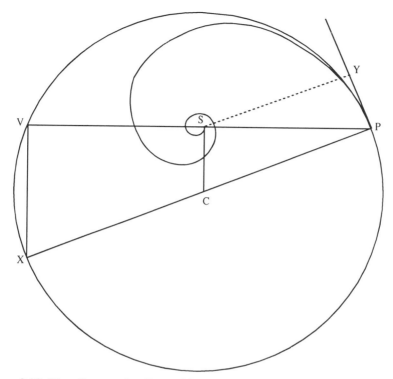

Figure 9.13 The diagram for Proposition 9 with the spiral extended from the point P into its pole S and the circle of curvature PVX displayed.

through the center of force SY has been added. Moreover, the line of force PS has been extended to a point V, where the distance PV is the chord of the circle of curvature at point P. This relationship is made explicit in the extended version of the diagram shown in figure 9.13, where PVX is the circle of curvature at the point P of the spiral whose pole is at the center of force S. Note that the chord of curvature PV through the pole S is bisected by the pole.

Step 1. Find "SY." Note from the figure that the angle SPY is the given constant angle of the spiral. Thus, $SY = PS \cos(SPY)$ and therefore SY and SP are "in a given ratio."

Step 2. Find "PV." As has been argued in the discussion of the initial solution in chapter 8, Newton uses without demonstration the relationship that the chord PV of the circle of curvature drawn through the pole of the spiral is equal to twice the pole distance SP. Thus, $PV = 2\ SP$ and therefore PV and SP are "in a given ratio."

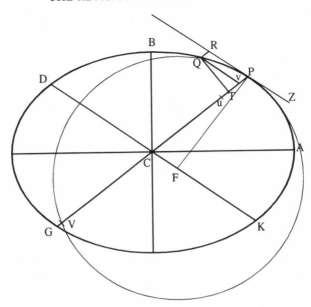

Figure 9.14 A revision of Newton's diagram for Proposition 10 in the 1713 *Principia* with the circle of curvature *PV* added.

Step 3. Find the circular dynamics ratio. Substituting *SY* and *PV* from Steps 1 and 2 directly into the circular dynamics ratio $1 / SY^2 \times PV$, the force is found to be proportional to $1/SP^3$.

THE REVISION OF PROPOSITION 10: AN ALTERNATE SOLUTION

The first major test of the revised dynamics ratio is set out in Proposition 10: an elliptical orbit with the force center located at the center of the ellipse. Figure 9.14 is the diagram that accompanies Proposition 10 in the revised editions. In contrast to the 1687 edition, separate diagrams are provided in the revised edition for Proposition 10 and in Proposition 11. The important change to note is the addition of the letter *V* in the lower left-hand section, which defines the chord of the circle of curvature through the center of the ellipse. The circle shown in figure 9.14 does not appear explicitly in Newton's diagram.

Step 1. Find "SY." Note in the diagram that the perpendicular *SY* that is required is not explicitly shown. In this example, however, the line *SY* is equal to the line *PF*, which is constructed perpendicular to the conjugate diameter *DK* and hence perpendicular to the tangent *PR*.[19]

Step 2. Find "PV." The line *PV* is the chord of the circle of curvature to be employed in the circular solution. The first nine lines of the circular solution are devoted to demonstrating the following relationship:

$$PV = 2CD^2 / PC$$

which is a property of an ellipse, where *PV* is the chord of the circle of curvature drawn through the center of the ellipse.[20]

Step 3. Find "the ratio." Thus, the two elements *PV* and *SY* that are required for the circular solution are known, and the force *F* is given as follows:

$$F \propto 1 / SY^2 \times PV = 1 / PF^2 \times 2(CD^2 / PC) = PC / 2(PF^2 \times CD^2) \propto PC$$

where *PF* x *CD* is a circumscribed area that is a constant for any ellipse (Lemma 12). Hence, the force *F* is directly proportional to the distance *PC*.

THE REVISION OF PROPOSITION 11: ALTERNATE SOLUTIONS

Newton does not employ the same technique for the alternate solution of the direct problem in Proposition 11 as he did in Proposition 10. This difference is made manifest in the diagram that accompanies Proposition 11 in the revised editions. The diagram for Proposition 10 was revised, but the diagram for Proposition 11 is almost identical to the diagram given in the first edition.[21] Note, therefore, that the points *V* and *Y*, which appear in the diagrams for Propositions 9 and 10, are missing for this diagram. Thus, the perpendicular *SY* and the chord of the circle of curvature *PV* are not indicated, either explicitly or implicitly. Figure 9.15 is a diagram for Proposition 11 that does show the normal *SY* and the chord *PV*. The direct Kepler problem that opens Section 3 was *not* solved following the paradigm established in Proposition 10 but it *could* have been, as it was in the unpublished proposed revisions. The following is such a solution:

An Unpublished Alternate Solution

Step 1. Find "SY." From the similar triangles *YSP* and *FPE* one has the following:

$$SY = SP (PF / PE)$$

Step 2. Find "PV." From a similar proof to that which appears in Proposition 10, the chord *PV* of the circle of curvature that passes through a focus of an ellipse (in contrast to passing through a center) is as follows:

$$PV = 2CD^2 / PE$$

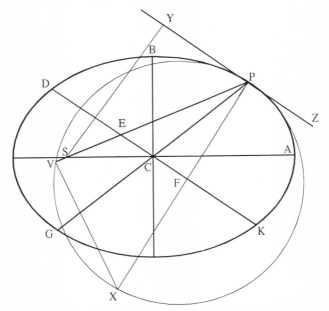

Figure 9.15 A version of the diagram for Proposition 11 with the circle of curvature *PVX* and the perpendicular *YS* added.

Newton obtained this result in the proposed revisions that were discussed in chapter 8.

Step 3. Find "the ratio." Thus, again the two elements *PV* and *SP* are known and the force *F* is given as follows:

$$F \propto 1 / (SY^2 \times PV) = PE^3 / (SP^2 \times 2CD^2 \times PF^2)$$

And from the additional knowledge that *PE* = *CA* (a constant), from the relationships in Lemma 12 (*PF* = *CA* x *CB* / *CD*), and from the definition of the constant *latus rectum L* of an ellipse ($L = 2CB^2 / CA$), the force is given as inversely proportional to the square of the line of force *SP*.

$$F \propto 1 / (SY^2 \times PV) \propto 1 / (L \times SP^2) \propto 1 / SP^2$$

But Newton did *not* elect to publish this solution to the Kepler problem, even though it is clear in his unpublished revisions of the 1690s that he could produce such a solution. Instead, he elected to use the comparison theorem that appears in Corollary 3 of Proposition 7 rather than the direct application of the circular dynamics ratio just given. The statement of the alternate method that appears in the revised editions is as follows:

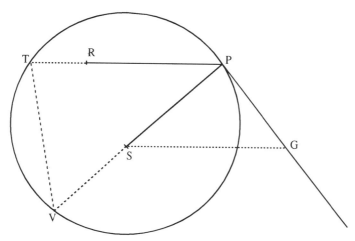

Figure 9.16 Based on Newton's figure for Corollary 2 of Proposition 7 in the revised *Principia*.

The Published Alternate Solution

Proposition 11. *Being directed to the center of an ellipse, a force by which a body* P *can be revolved on that ellipse would be* (*by Proposition 10, Corollary 1*) *as the distance* CP *of the body from the center* C *of the ellipse; let* CE *be drawn parallel to the tangent* PR *of the ellipse; and the force by which the same body* P *can be revolved around any other point* S *of the ellipse, if* CE *and* PS *should meet at* E, *will be as* PE³ / SP² (*by Proposition 7, Corollary 3*); *that is if point* S *should be the focus of the ellipse, and thus* PE *should be given,* [*then*] *the force will be as* SP² *reciprocally. Which was to be found.*

Figure 9.16 is the drawing that accompanies Corollary 2 of Proposition 7 in which Newton demonstrated that the ratio of the forces F_S / F_R (forces that are directed to any two different points S and R, and that will maintain the same body moving along the same circle *PTV*) is given by the comparison dynamics ratio (i.e., $F_S / F_R = SP \times RP^2 / SG^3$, where SG is parallel to *RP*). In Corollary 3, Newton extended the results given in Corollary 2 for a "circular orbit" to "any orbit" simply by noting in the final sentence, "For the force in this orbit at any point *P* is the same as in a circle of the same curvature."

Figure 9.17 shows a portion of the drawing that Newton provided for Proposition 11 with my addition of the line *CG'* corresponding to the line *SG* in the diagram for Corollary 3 in figure 9.16. The two general points *S* and *R* from the diagram for Corollary 3 become the points for the center of the ellipse *C* and the focus of the ellipse *S* (i.e., $F_C / F_S = CP \times SP^2 /$

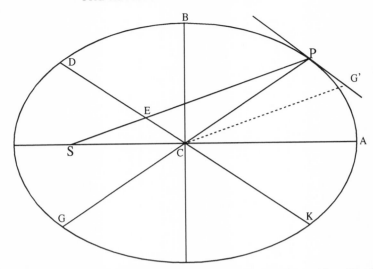

Figure 9.17 A restricted version of Newton's diagram for Proposition 11 with the line CG' from Corollary 2 of Proposition 7 added.

CG^3), where S (in Proposition 7) becomes C (in Proposition 11) and R (in Proposition 7) becomes S (in Proposition 11). In Proposition 10, which precedes the Kepler problem, Newton has determined that the force F_C required to maintain an elliptical orbit, when the force is directed to the center C of the ellipse, is given by $F_C \propto CP$. Substitution of that result into the comparison dynamics ratio gives the force to the focal point S as $F_S \propto CG^3 / SP^2$. From the figure, $CG = PE$ because the diameter DK is parallel to the tangent PG. Further, Newton has demonstrated earlier in Proposition 11 that $PE = AC$, where $2AC$ is the major axis, a constant of the ellipse. Thus, $F_S \propto 1 / SP^2$, as required. It is an efficient solution but one that leads the reader on a tortuous trail as he or she traces back through Proposition 7 to its first principles in Proposition 6, Theorem 5. When Newton completes this solution, he makes the following observation at the close of Proposition 11:

> [Proposition 11] *With the same brevity with which we reduced the fifth Problem to the parabola, and hyperbola, we might do the like here: But because of the dignity of the problem and its use in what follows, I shall confirm the other cases by particular demonstrations.*

Newton is not as expansive, however, in his efficient but indirect presentation of the comparison theorem as he is in the extended presentation of the Kepler problem. The reasons for his choice of this restricted type of presentation in the published editions lie buried in his unpublished revisions. There he demonstrated each of his three methods of solution in a new and independent proposition. Here he must fold the three methods into the existing structure of the first edition of the *Principia*.

CONCLUSION

Thus, the reader of the revised published editions of the *Principia* is presented with new insights into the basic dynamics of planetary motion, but that information is hidden in a labyrinth not present in the first edition. In the first edition, the basic linear dynamics ratio is developed in Proposition 6 and applied in sequence to a series of five problems, each problem presented in a separate proposition and clearly labeled as a problem. In the revised published editions, Proposition 6 contains two basic dynamics ratios that provide alternate solutions for the first four of the exemplar problems that follow it. In Proposition 7, the direct problem of a circular orbit with a center of force on the circumference of the first edition is extended in its published revised format to include any general force center for a circular orbit; then it is extended to provide the ratio of the forces for two different centers; and then that result is further extended in the final corollary of the problem to include any orbit whatsoever (i.e., the comparison theorem). In the unpublished revision, this ratio is the premier theorem and is set forth clearly as the primary method of solving all the problems of the conic sections. In the published revised editions, it is hidden away as a corollary to a problem in Proposition 7.

In the unpublished revisions, the new Lemma 12 provides the basis for employing the chords of the circles of curvature to ellipses to produce a general dynamics ratio for ellipses from the circular dynamics ratio. Further, it gives the specific chords of the circle of curvature required in the solution of the specific problems of focal and central force centers for ellipses. In the published revised editions the problem of the central ellipse (Proposition 10) is solved in this fashion but the alternate solution for the problem of the focal ellipse (Proposition 11) is not solved by the logical extension of the technique. Rather, the alternate solution is produced using the comparison ratio from Proposition 7 and the alternate solution for a central ellipse from Proposition 6.

Why did Newton give up the eloquent and graceful development of the basic dynamics of the unpublished revisions for the patchwork labyrinth of the published revisions? Rather than three complete and complementary sets of solutions for the direct problems of circular and elliptical motion with their fundamental theorems clearly set forth, the published revision presents a mixture of solutions with theorems appearing in problem solutions of other theorems. In discussing a proposed revision of all the lemmas to accompany the proposed revision of the propositions, Whiteside states that "inertia prevailed and in the revised second edition (and in all subsequent editions thereafter) the lemmas of the 1687 edition of the *Principia* all reappear unchanged in location and with only minimal alterations in their verbal text."[22] And, presumably, the same "inertia" explains Newton's failure to revise the location and description of the propositions. It would have been a Herculean task to seek out all the references to earlier lemmas and propositions, revise them, and then renumber them. Rather, he contented himself with tucking pieces of the new theorems and solutions into whatever existing nooks and crannies he could find. And thus, the general comparison theorem appears as a corollary of a problem on circular motion and not as an independent theorem.

It is regrettable that the full revision was not done. The published primary solution to the direct Kepler problem cloaks its linear dynamics in a heavy overlay of mathematics, and the revised published alternate solution wanders back through the labyrinth of the hidden comparison ratio in Proposition 7 and ultimately to the revised Proposition 6. In contrast, the three unpublished solutions to the direct Kepler problem follow in a clear and simple fashion from three separate and clearly identified dynamics theorems.

TEN

Newton's Dynamics in Modern Mathematical Dress

The Orbital Equation and the Dynamics Ratios

Throughout this book I have attempted to view Newton's creative process in terms of the dynamics and mathematics that preceded his analysis, rather than to view it with hindsight from a modern perspective. In this closing chapter, however, I reverse the procedure and express Newton's dynamic measures of force in current mathematical notation. Contemporary textbooks in physics present a second-order differential equation called "Newton's Second Law" in the familiar form of $F = ma$. For motion in one dimension x, this equation has only one component equation: $F(x) = m(d^2x/dt^2)$. The force function $F(x)$ is equal to the product of the mass m and the second derivative of distance x with respect to the time t (i.e., the acceleration (d^2x/dt^2)). For motion in two dimensions, there must be two such equations: one in x and one in y. Motion under a force directed toward a fixed center, such as gravitational force acting toward the sun, is confined to a plane and thus requires only two such equations.[1] If these equations are expressed in terms of the polar coordinates (radius r and angle θ) instead of the Cartesian coordinates (x and y), then the two component equations can be written as follows:[2]

$$F_r = m[(d^2r/dt^2) - r(d\theta/dt)^2] = F(r)$$

$$F_\theta = m[r(d^2\theta/dt^2) + 2(dr/dt)(d\theta/dt)] = 0$$

The force in the radial direction F_r is given as a function $F(r)$ of the polar radius r alone. The angular force F_θ is zero, because the force is directed only toward the center along the radius r and thus it has no angular component.

Given the force function F_r, one can solve the equations of motion for the path of the particle as a function of time. It is possible, however, to eliminate time t as a parameter in these two equations and to write the following expression for the force in terms of the radius r and the angle θ: the *polar orbital equation*.

$$F_r = K^2 r^{-2} [r^{-1} + d^2(r^{-1}) / d\theta^2]$$

where K is a constant, and the derivative is expressed in the more compact form in terms of r^{-1} (i.e., the inverse of the radius).[3] Thus, if the path of the particle $r = r(\theta)$ is given (i.e., if the polar radius r is known as a function of the polar angle θ), then the force F_r can be found as a function of the radius r simply by taking the second derivative $d^2(r^{-1})/d\theta^2$ and substituting it into the polar orbital equation. Problems of this sort—find the force from a given path and center of force—are the direct problems that Newton solved in the opening sections of the *Principia*.

As an example of the application of the polar orbital equation, consider the direct problem given in Proposition 9 of the *Principia*: find the force required to maintain an orbit that is an equiangular spiral with the center of force located at the pole of the spiral. The equation of the spiral path is given by $r = r(\theta) = Ae^\theta$, where A is a constant, and the reciprocal r^{-1} is $Ae^{-\theta}$. The second derivative of $Ae^{-\theta}$ with respect to θ is simply $Ae^{-\theta}$, which is equal to r^{-1}, and the polar orbital equation gives the following functional dependence of the force F_r on the radius r:

$$F_r = K^2 r^{-2} [r^{-1} + d^2(r^{-1}) / d\theta^2] = K r^{-2} [r^{-1} + r^{-1}] = 2K r^{-3}$$

Thus, as Newton has demonstrated in his solution to the spiral/pole direct problem in Proposition 9, the force is inversely proportional to the cube of the radius.

The solution to the distinguished Kepler ellipse/focus problem of Proposition 11 can be solved in a similar fashion. The equation of an ellipse relative to an origin fixed in a focus is given by $r^{-1} = [A + B \cos(\theta)]$ and the second derivative $d^2(r^{-1}) / d\theta^2$ is simply $- B \cos(\theta)$. Thus, the polar orbital equation gives the following functional dependence of the force F_r on the radius r:

$$F_r = K^2 r^{-2} [r^{-1} + d^2(r^{-1}) / d\theta^2] = K^2 r^{-2} [(A + B \cos(\theta)) - (B \cos(\theta))]$$

$$= (K^2 A) r^{-2}$$

The force F_r is inversely proportional to the square of the radius r, as Newton has demonstrated in his solution to this problem in Proposition 11. If one has mastered a first course in calculus, then the solution of the polar orbital equation appears to be much simpler than Newton's solution in Proposition 11 using the linear dynamics ratio. That apparent simplicity is deceptive, however, for whereas Newton's analysis clearly sets out the basic principles in terms of the parabolic approximation, much of the dynamics of the orbital solution is hidden in the algorithms of the calculus.

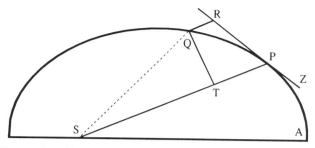

Figure 10.1 Based on Newton's diagram for Proposition 6 of the 1687 *Principia*.

THE ORBITAL EQUATION AND CURVATURE

The *polar* orbital equation can be written in an alternate form as the *curvature* orbital equation, $F_r = K^2 / (r^2 \rho \sin^3(\alpha))$, where ρ is the radius of curvature and α is the angle between the tangent to the curve and the radius r. The demonstration of the equivalence of the two equations is based upon three analytical elements: Newton's demonstration of Kepler's area law, Newton's circular approximation, and Newton's original work on curvature.

The first relationship is an expression of Newton's Proposition 1, Kepler's law of equal areas in equal times. Figure 10.1 is based upon the diagram for Proposition 6 in the 1687 edition of the *Principia*. The general curve is *APQ*, the tangent to the curve is *ZPR*, and the center of force is *S*. The area A of the triangle *SPQ* is equal to $(\tfrac{1}{2}) SP \times QT$, or what is equivalent, $(\tfrac{1}{2}) SP \times PR \sin(\alpha)$, where α is the angle *SPR* between the radius *SP* and the tangent *ZPR*. If the radius *SP* is written as r and the tangential velocity at *P* as v, then in a given time Δt, the line segment *PR* is equal to $v \Delta t$, and the area A is given by $r v \Delta t \sin(\alpha)$. Thus, the rate at which twice the area A is swept out is given as follows:

$$2\Delta A / \Delta t = (r v \Delta t \sin(\alpha)) / \Delta t = r v \sin(\alpha) = K$$

where K is a constant proportional to the area swept out per unit time. This relationship is mathematically equivalent to the modern law of conservation of angular momentum, where K is the angular momentum per unit mass.

The second relationship is an expression of Newton's circular approximation: the replacement of motion along an incremental arc of the general curve at point *P* with uniform circular motion along an incremental arc of the circle of curvature at the same point. Figure 10.2 is based upon the diagram for Proposition 6 in the 1713 edition of the *Principia*. In his revised diagram, Newton added the normal to the tangent through the center of force *YS* and extended the line *PS* to the point *V* to display the

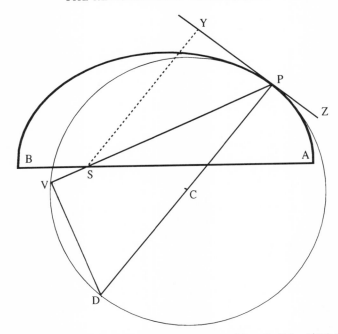

Figure 10.2 Based on Newton's diagram for Proposition 6 of the 1713 *Principia*.

chord of curvature through the center of force S. I have removed the points Q, R, and T, and I have explicitly displayed the circle of curvature PVD at the point P, where the center of the circle of curvature is at C and the diameter of curvature is PD (i.e., 2ρ). The angle PVD between the chord and diameter of the circle is a right angle and α is the angle SPY (equal to the angle SPR in figure 10.1). The component of force F_C directed toward the center of curvature C is given by

$$F_C = F_S \sin(\alpha) = v^2 / \rho$$

where F_S is the force per unit mass directed toward the center of force S, and ρ is the radius of curvature. The force F_C directed toward the center of curvature C provides the centripetal circular acceleration v^2 / ρ as required in Proposition 4 of the *Principia*. Newton has explicitly employed this uniform circular replacement in Lemma 11, Corollary 3 and in Proposition 7, Corollary 5; and he has implicitly employed it in other discussions. Combining the two relationships, the force F_S directed toward the center of force S can be written as:

$$F_S = K^2 / (r^2 \rho \sin^3(\alpha))$$

The third relationship concerns Newton's original work on curvature, which was extensive. It contains in detail what appears in the *Principia* only in outline. Central to the analysis is Newton's expression for the radius of curvature ρ in polar coordinates (r, θ), which he developed in the early 1670s. Newton expressed it in the following form:[4]

$$\rho = r(1 + z^2)^{3/2} / [(1 + z^2) - z']$$

where z' is $dz/d\theta$ and z is the slope of the curve $(1/r)(dr/d\theta) = \text{ctn}(\alpha)$. It can be demonstrated that $(1 + z^2)^{3/2} = \sin^{-3}(\alpha)$, where α is the angle between the radius r and the tangent to the curve, and that $[(1 + z^2) - z'] = r(r^{-1} + d^2(r^{-1})/d\theta^2)$.[5] Thus, the expression for the radius of curvature ρ can be written also as follows:

$$\rho = 1/[\sin^3(\alpha)(r^{-1} + d^2(r^{-1})/d\theta^2)]$$

If this expression is solved for $(r^{-1} + d^2(r^{-1})/d\theta^2)$, then the *polar* orbital equation can be written in an alternate form as a function of r, ρ, and α: the *curvature* orbital equation.

$$F_r = K^2 r^{-2}(r^{-1} + d^2(r^{-1})/d\theta^2) = K^2 / (r^2 \rho \sin^3(\alpha))$$

The curvature orbital equation, expressed in terms of the radius r, the radius of curvature ρ, and the angle between the tangent and the radius α (i.e., $F_r = K^2/(r^2 \rho \sin^3(\alpha))$), echoes Newton's cryptic statement of 1664, in which he argued that the force (F_r) for elliptical motion at a point (r) can be found by the curvature (ρ) at that point, if the motion (v and α) at that point is given (i.e., $F_s = v^2/(\rho \sin(\alpha))$).[6]

THE DYNAMICS RATIOS AND THE ORBITAL EQUATION

The linear dynamics ratio $QR/(QT^2 \times SP^2)$, which Newton derives in Proposition 6, is the functional equivalent of the contemporary orbital measure. Whiteside has demonstrated the relationship by expanding the geometric terms SP, QT, and QR in a power series, holding terms to the second order of the differential angle $d\theta$, and expressing them as a function of the analytic terms r, θ, and their derivatives.[7] It is also possible to use Lemma 11 to demonstrate the same result. Figure 10.3 is a revision of the diagram for Lemma 11 into the form that applies to Proposition 6. Lemma 11 states that $AB^2/BD = AG$ and if B approaches A, then AG approaches AJ, the diameter of the circle of curvature 2ρ. Written in terms of the elements of Proposition 6, $AB \rightarrow QT/\sin(\alpha)$, $BD \rightarrow QR \times \sin(\alpha)$, and $AG \rightarrow 2\rho$. Thus, the limiting value of the ratio AB^2/BD is given by the following expression:

$$AB^2/BD \rightarrow (QT/\sin(\alpha))^2/(QR \times \sin(\alpha)) = QT^2/(QR \times \sin(\alpha)^3) = 2\rho$$

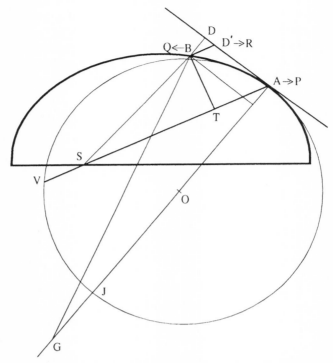

Figure 10.3 Newton's diagram for Lemma 11 adapted to the diagram for Proposition 6.

which may be solved for the discriminate $QR / QT^2 = 1 / (2\rho \sin^3(\alpha))$. Thus, the linear dynamics ratio $QR / (QT^2 \times SP^2)$ is equal to twice the curvature orbital measure $1 / (r^2 \rho \sin^3(\alpha))$.

The circular dynamics ratio $1 / (SY^2 \times PV)$, which Newton introduced into the revised edition, also is identical with the curvature orbital measure. From figure 10.2 (given previously), the angle SPY is equal to the angle PDV or α (in fig. 10.2, both YS and PD are normal to the tangent YPZ and the angles PVD and YPD are right angles). The chord of the circle of curvature through the center of force PV is equal to $PD \sin(\alpha)$ or $2\rho \sin(\alpha)$ and the normal to the tangent SY is equal to $SP \sin(\alpha)$ or $r \sin(\alpha)$. Thus, the circular dynamics ratio $1 / (SY^2 \times PV)$ also is equal to twice the curvature orbital measure $1 / [r^2 \rho \sin^3(\alpha)]$.[8]

If the force is expressed in terms of the curvature orbital measure, then the solutions to the direct problems that Newton selected for the 1687 *Principia* fall into an interesting pattern.

Proposition	Orbit/Force Center	$\sin(\alpha)$	Radius of Curvature	Force $\propto 1 / (SP^2 \rho \sin^3(\alpha))$
7	circle/circumference	$SP/2\rho$	ρ = constant	Force $\propto 1/SP^5$
9	spiral/pole	α = constant	$\rho = SP/\cos(\alpha)$	Force $\propto 1/SP^3$
10	ellipse/center	PF/SP	$\rho = CD^2 / (SP \times \sin(\alpha))$	Force $\propto SP$
11	ellipse/focus	PF/PE	$\rho = CD^2 / (PE \times \sin(\alpha))$	Force $\propto 1/SP^2$

For Propositions 7 and 9, the solution is simplicity itself because ρ and α are respectively constant. For both Propositions 10 and 11, however, one must note that the product $PF^2 \times CD^2$ is proportional to the area of the ellipse, a constant, and for Proposition 11 that PE is equal to the semi-major axis of the ellipse, also a constant.

Proposition 7. $SP^2 \rho \sin^3(\alpha) = SP^2 \times \rho \times (SP/2\rho)^3$
$= SP^5 / 2\rho^2 \propto SP^5$

Proposition 9. $SP^2 \rho \sin^3(\alpha) = SP^2 \times (SP/\cos(\alpha)) \times \sin^3(\alpha) \propto SP^3$

Proposition 10. $SP^2 \rho \sin^3(\alpha) = SP^2 (CD^2 / SP \times SP/PF) (PF/SP)^3$
$= (PF^2 \times CD^2) / SP \propto 1/SP$

Proposition 11. $SP^2 \rho \sin^3(\alpha) = SP^2 (CD^2 / PE \times PE/PF) (PF/PE)^3$
$= SP^2 (PF^2 \times CD^2) / PE^3 \propto SP^2$

The choice of the orbit and focal center for Proposition 11 in 1684 was dictated by the physical problem of the planets, but the choice of the other examples was arbitrary. They may have been suggested by the work on curvature, which Newton began as early as 1664.[9] In his *Methods of Series and Fluxions* of 1671, Newton calculated the radius of curvature for a number of examples, including conic sections and spirals.[10] Newton's statement of 1664 indicated that he intended to use his work on curvature to solve direct problems of orbital motion. Whatever method he had in mind when he made the statement in 1664, however, it may not have been in the form of the curvature orbital measure $1/(r^2 \rho \sin^3(\alpha))$, because that measure entails the area law, which Newton did not discover before 1679.

Michael Nauenberg has suggested that before the discovery of the area law Newton could have employed a numerical method based on curvature to evaluate orbits. Nauenberg demonstrates an iterative computational method to reproduce the figure for the orbit of a body subject to a constant central force such as Newton sent to Hooke in 1679.[11] Moreover, Nauenberg suggests that Newton could have used the material in his 1671 *Methods of Series and Fluxions* to produce an analytical measure of the force, independent of the area law, to solve direct problems.[12] There is no evidence that Newton actually carried out such calculations, but Nauenberg argues that there is considerable circumstantial evidence. For example, in the same letter of 13 December 1679 in which Newton sent Hooke the drawing of an orbit obtained by a numerical method for a constant force, Newton also noted that if the force increased as the distance decreased, then the body may "by an infinite number of spiral revolutions descend continually till it cross the center."[13] Nauenberg suggests that Newton knew that the reciprocal cube force generated the constant angle spiral (logarithmic spiral). He notes that Newton's observation of "an infinite number of spiral revolutions" into the center cannot be deduced from a numerical solution of orbital motion, because that technique can provide only a *finite* number of revolutions in approaching the center. Nauenberg also points to Newton's choice of the constant angle spiral produced by a reciprocal cube force directed to the pole of the spiral as an example in the *Principia* for both the spiral/pole direct problem in Proposition 9 and the reciprocal cube inverse problem in Proposition 31.

> It is also noteworthy that later in the *Principia,* Newton uses as an example the $1/r^3$ force law, rather than the physically more interesting $1/r^2$ case, to solve explicitly the inverse problem, given the force law obtain the orbit (Book One, Proposition 31, Corollary 3).[14]

Nauenberg argues that Newton could have solved various direct problems before 1679, and he challenges the received opinion that Newton could not have done so until after his discovery of the area law in 1679. In addition to this new debate over a solution for the direct problem, there is a long-standing debate over the outline for a solution to the inverse problem that Newton added to the 1713 *Principia* and extended in 1726.

THE REVISION OF COROLLARY 1 OF PROPOSITION 13

In Corollary 1 of Proposition 13 in the 1687 *Principia*. Newton assumed that the solutions to the conic/focus direct problems given in Proposition 11, 12, and 13 also constituted a solution to the reciprocal square inverse problem.

> *From the last three propositions it follows that if any body* P *should depart from position* P *along any straight line* PR, *with any velocity, and is at the same time acted upon by a centripetal force that is reciprocally proportional to the square of the distance from the center,* [*then*] *this body will be moved in one of the sections of conics having a focus at the center of forces; and conversely.*

This statement is an improvement over the initial version in *On Motion* sent to Halley in 1684, in which only the solution to the ellipse/focus direct problem was presented and the other two conic sections were not discussed. Nevertheless, even as it stands in the 1687 *Principia*, this statement is unsatisfactory, and was criticized by Johann Bernoulli before the publication of the 1713 *Principia*.[15] Newton's solution to the direct problem of conic/focal motion in the "last three propositions" (i.e., Propositions 11, 12, and 13) does not by itself constitute a solution for the reciprocal square force inverse problem. As Whiteside puts it,

> The hidden assumption here made that no curve other than a conic may, in an inverse-square force-field centered on *S*, satisfy all possibilities of motion at *P* wants—for all its manifest plausibility—an explicit, rigorous justification, and Newton was later fairly criticized by Johann Bernoulli for merely presupposing its truth without demonstration.[16]

Newton himself recognized the need to justify his assumption in the 1687 *Principia* of a solution to the inverse problem. In 1709, as the revised edition of the *Principia* was being prepared, Newton wrote to his editor Roger Cotes and requested that he add the following statement to Corollary 1 of Proposition 13 (Newton added the words in brackets to the 1726 *Principia*).

> Corollary 1. *And the contrary. For the focus, the point of contact, and the position of the tangent being given, a conic section may be described, which at that point shall have a given curvature. But the curvature is given from the given centripetal force [and the body's velocity]: and two orbits mutually touching one the other, cannot be described by the same centripetal force [and the same velocity]*.[17]

Newton intended this extension to serve as an outline for the solution to the inverse problem, and evidently it satisfied Johann Bernoulli, for in 1719 he wrote to Newton as follows:

> Gladly I believe what you say about the addition to Corollary 1, Proposition 13, Book One of your incomparable work, the *Principia*, that this was certainly done before these disputes began, nor have I any doubts that the demonstration of the inverse proposition, which you have merely stated in the first edition of the work, was yours; I only said something against the form of that assertion, and wished that someone would give an analysis that led *a priori* to the truth of the inverse [proposition] and without supposing the direct [proposition] to be already known. This indeed, which I would not have said to your displeasure, I think was first put forward by me, at least so far as I know at present.[18]

Whiteside argues that Newton could have employed a general polar curve with given curvature to produce a solution for the reciprocal square force inverse problem without assuming the solution to the conic/focal direct problem.[19] Newton did not, however, explicitly produce that solution (as he did for the reciprocal cube inverse problem in Proposition 31). Moreover, the validity of the outline of the solution of the inverse problem given in Corollary 1 of Proposition 13 in the 1726 *Principia*, which does assume the solution to the direct problem, continues to receive an occasional challenge. But now, as in the past, every challenger produces a number of defenders. One recent challenger has claimed that Newton's solution, even as given in outline in the corollary, is radically flawed and contains a gross, irreparable fallacy. Another critic allowed that a gap may exist in the logic of Newton's outline, but argued that the gap is intuitively easy to fill. Other defenders argued that no gap of any sort exists.[20] The Russian mathematician V. I. Arnol'd, however, saw no real basis for such a discussion. He argued, "The spirit of modern mathematics has penetrated to a number of physicists . . . and they have begun to worry about questions that earlier nobody would have talked about seriously." He arrives at the following conclusion:

> In fact, all this argument is based on a profound delusion. Modern mathematicians actually distinguish existence theorems and uniqueness theorems for differential equations and even give examples of equations for which the existence theorem is satisfied but the uniqueness theorem is not. . . . Thus, in general, uniqueness does not follow from the existence of a solution, but everything will be in order if the solution produced depends smoothly on the initial condition. . . . For each initial condition [Proposition 17] Newton produced a solution, described it, and from this description it became obvious straight away that the solution depends smoothly on the initial condition. . . . Of course, one could raise the objection that Newton did not know this theorem. . . . But he certainly knew it in essence.[21]

It is of interest to note that in the outlined solution of Corollary 1 of Proposition 13, as elsewhere in the *Principia*, Newton assumed on the part of the reader a background in the mathematics of curvature. In Lemma 11, the reference to curvature appeared only in a parenthetical expression, even though curvature was primary to the relationship developed in the lemma. In the final corollary to Proposition 7, the extension of the analysis on circular motion to the comparison theorem and general motion was defended by a final single sentence that called upon curvature. In Proposition 13, Newton again called upon his extensive work on curvature without an elaboration.

A DETAILED SOLUTION OF THE INVERSE PROBLEM

I now produce a solution to the inverse problem using Newton's equation for curvature that does not explicitly employ the solution to the direct problem (i.e., Propositions 11, 12, and 13).[22] Instead, the solution employs Newton's work on curvature expressed in contemporary notation. There is no evidence that Newton actually produced such a solution for the inverse problem for gravitational force, but Whiteside argues that he was capable of doing so.[23] The expression for the force obtained from the area law and the circular approximation is given by the curvature orbital equation as follows:

$$F_S = K^2 / (r^2 \rho \sin^3(\alpha))$$

Newton's original expression for $\rho = r(1 + z^2)^{3/2} / [(1 + z^2) - z']$ can be expressed in terms of $(1 + z^2)$ as follows:[24]

$$\rho = r(1 + z^2)^{3/2} / [(1 + z^2) - (r/2) d(1 + z^2) / dr]$$

Moreover, I have demonstrated that $\sin^3(\alpha) = (1 + z^2)^{3/2}$. If ρ and $\sin(\alpha)$ are written in terms of $(1 + z^2)$, then the expression for the force F_S can be rearranged and expressed as follows:

$$d(1 + z^2) / dr - 2(1 + z^2) / r = -(2 / K^2)(r^2 F)$$

If the force F is given as c / r^2, and $(1 + z^2)$ is written as the function $f(r)$, then the equation reduces to:

$$df / dr - 2 f/r = -2A$$

where $A = c/K^2$. This equation is a first-order linear differential equation whose complementary solution f_c satisfies the equation $df_c / dr - 2f_c / r = 0$ and is given by $f_c = Cr^2$, where C is an arbitrary constant. The particular solution, f_p, is given by $2Ar$, and thus the full solution f is:

$$f = f_c + f_p = 2Ar + Cr^2$$

Substituting $(1 + z^2)$ for f (as defined) and $(B^2 - A^2)$ for C (where B is arbitrary and A is given), and solving for z (defined as $(r^{-1}) dr / d\theta$), the following relationship results:

$$z = r[B^2 - (r^{-1} - A)^2]^{1/2} = (r^{-1}) dr / d\theta$$

Solving for the differential angle $d\theta$ and integrating, one obtains the following relationship:

$$\theta = \int [(r^2)(B^2 - (r^{-1} - A)^2)^{1/2}]^{-1} \, dr = \cos^{-1}[(r^{-1} - A) / B] - \varepsilon$$

where ε is a constant of integration. If that equation is solved for r^{-1}, then the polar equation of the general conic is given as follows:

$$r^{-1} = A + B \cos(\theta + \varepsilon)$$

Thus, as Newton argued in the statement he added to Proposition 13 in the revised *Principia*, the path is uniquely determined given the initial position and the curvature from the force and velocity.

CONCLUSION

Newton's contribution to dynamics must be measured not in terms of its correspondence with modern standards or modern methods, but rather in terms of the innovative and ingenious insights revealed in his initial analysis. As early as 1665, and certainly before 1669, Newton had laid the foundations for his mature mathematical and dynamical analysis. Both the polygonal approximation and the parabolic approximation appeared in his pre-1669 analysis of uniform circular motion, and they were carried forward into his analysis of noncircular motion after 1679 following his demonstration of Kepler's area law. The circular approximation used in the alternate solutions of the 1713 *Principia* was based upon his work in 1664 on curvature. The solution to the direct Kepler problem, which is the hallmark of the 1687 *Principia*, had its roots in his statement of 1664.

> *If the body* b *moved in an Ellipsis then its force in each point (if its motion in that point bee given) may bee found by a tangent circle of Equall crookednesse with that point of the Ellipsis.*[25]

That statement reached its fruition in the curvature orbital measure of force $1 / (r^2 \rho \sin^3(\alpha))$, which has been demonstrated to be an alternate form of Newton's linear dynamics ratio of the 1687 *Principia*, his circular dynamics ratio of the 1713 *Principia*, and the modern polar orbital equation. If any single measure deserves the title of the key to Newton's dynamics, it is the curvature measure. To appreciate Newton's work on dynamics fully is to appreciate how it began in 1664 in the *Waste Book*, how it matured in the *Principia*, and how it relates to contemporary expressions of dynamics.

APPENDIX

Bust of Mary Somerville, by Chantrey, currently located in the library of the Royal Society in London. Somerville died in 1872, at the age of 92, after an active career as a scientific writer. During her lifetime women were not eligible for membership, but the Royal Society did recognize her achievements. In 1832, sixty-four subscribers, headed by the Duke of Sussex, pledged funds to have this bust made and placed in the Royal Society. In 1922, it was decided that in consequence of legislation enacted since the war of 1914–1918 women were eligible for election. The first two women Fellows were elected at the close of the war of 1939–1945. Copyright © The Royal Society. Reproduced by permission.

An English Translation of Sections 1, 2, and 3 of Book One from the First (1687) Edition of Newton's *Mathematical Principles of Natural Philosophy*

In the first edition of the *Principia,* Newton advised his readers to concentrate on the first three sections of Book One and then to select topics of interest from Book Three. This advice remained unchanged in the two revised editions published during his lifetime and has guided my selection of text for translation. I have selected the first edition for translation, although it is the third and final edition on which most base their study of Newton's dynamics. I maintain, however, that such a study should begin not with the third edition and its many additions and revisions, but rather with the first edition and its relatively straightforward presentation. In the first edition, Newton clearly explicates his analysis with a single method applied consistently to several problems. Until the reader understands Newton's original method and his unpublished restructuring, Newton's additions to the much revised third edition distract rather than enrich the reader.

This Appendix contains a translation into English of the first three sections of Book One of the first Latin edition of the *Principia*. The translation of Newton's text from the Latin into English requires the skills of a classical linguist, a mathematician or physicist, and a historian of science; a combination rarely found in one individual. This translation was produced by combining the talents of a classical scholar and a physicist turned historian of science: Mary Ann Rossi and myself. She produced a first draft of a section, which I then read and commented upon, and we both discussed and revised the translation. My challenge was to produce a text that made mathematical and physical sense, and the challenge for her was to produce a text true to the Latin text. When successful, the translation conveys Newton's mathematical and physical intentions while remaining faithful to the Latin. The occasionally awkward language of the translation must be weighed against the virtue of a translation that is relatively free from the translator's bias.

Nevertheless, a faithful translation does not overcome all obstacles to a clear rendering of Newton's intentions. In Problems 1, 2, and 3 in *On Motion*, for example, Newton writes *corpus gyrat* (a body orbits). These same problems appear in the *Principia* as Propositions 7, 10, and 11 where Newton writes *gyretur corpus* (let a body be orbited) for the first two and *revolvatur corpus* (let a body be revolved) for the third. The former ("a body orbits") does not imply an external agent responsible for the motion, while the latter ("let a body be orbited") might be so construed.[1] No supporting evidence suggests, however, that Newton intended this change in the Latin to imply a change in physical doctrine. On the contrary, in Definition 8 Newton states that he considers "forces not physically, but only mathematically. Hence, let the reader beware lest, because of words of this kind [attraction or impulse], they should think either that I am defining . . . a physical cause or reason, or that I am attributing forces truly and physically to centers (which are mathematical points)." Still, Newton is consistent in his revision of the voice of the verb in all three problems while leaving the remainder of the text unchanged. Most translations of this text translate both forms as "a body orbits." Given our self-imposed constraints to follow the Latin text as closely as possible, we have made the distinction. So let the reader beware.[2]

The only liberty taken with the Latin original was to modify the format of the text in order to clarify each part of a proposition. For example, in the solution of the Kepler problem in Proposition 11, the description of the diagram is differentiated from the details of the demonstration by separating each into its own paragraph; the paragraph containing the demonstration in turn is separated into a series of logical steps by inserting a number of line breaks; and the conclusion is set forth in yet another paragraph. Newton presents all of the text (description, demonstration, and conclusion) in a single paragraph. I find, however, that there must be room for the eye and the mind to rest if the entire proposition is to be read with ease and so I have introduced the separations.

The following is an English translation, without commentary, of the first three sections of Book One of the 1687 edition of the *Principia*.[3] Section 1 contains the lemmas that provide a formal defense of the limiting procedures that Newton employs in the sections that follow. Section 2 contains the basic paradigm for the solution of direct problems, and examples of such solutions. Section 3 opens with the solution to the direct problem of elliptical/focal planetary motion (Proposition 11) and concludes with what Newton may have intended as the solution to the inverse problem (Proposition 17). The reader may wish to begin, however, by selecting topics of interest, such as the demonstration of Kepler's law of equal areas in equal times (Theorem 1 in *On Motion* and Proposition 1 in the *Principia*), or the derivation of the basic paradigm for the solution of direct problems (Theorem 3 in *On Motion* and Proposition 6 in the *Principia*), or perhaps the solution to the Kepler problem (Problem 3 in *On Motion* and Proposition 11 in the *Principia*). The content of these propositions in the 1687 *Principia* has changed very slightly from their presentation in *On Motion*, and the reader can refer to the detailed commentaries earlier in this book if questions arise. If there are questions on a lemma or proposition in the *Principia*, then the reader can use the following list to locate relevant earlier commentaries:

Translation	*Chapter*
Lemma 11	Chapter 7, Lemma 11
Proposition 1	Chapter 4, Theorem 1
Proposition 2	No reference
Proposition 3	No reference
Proposition 4	Chapter 4, Theorem 2
Proposition 5	No reference
Proposition 6	Chapter 4, Theorem 3
Proposition 7	Chapter 5, Problem 1
Proposition 8	No reference
Proposition 9	Chapter 7, Proposition 9
Proposition 10	Chapter 5, Problem 2
Proposition 11	Chapter 5, Problem 3
Proposition 12	Chapter 5, Problem 3
Proposition 13	Chapter 5, Problem 3
Proposition 14	No reference
Proposition 15	Chapter 6, Theorem 4
Proposition 16	No reference
Proposition 17	Chapter 6, Problem 4

MATHEMATICAL PRINCIPLES OF NATURAL PHILOSOPHY

Author, *ISAAC NEWTON, Trinity College, Cambridge.*
Lucasian Professor of Mathematics & Friend of the Royal Society

LICENSED BY
S. PEPYS, *Royal Society* PRESIDENT
July 5, 1686

LONDON
By the order of the *Royal Society*, printed by *Joseph Streater* and sold by several bookdealers. *In the year* 1687.

PREFACE
TO THE READER

Since the Ancients (as Pappus *tells us) regarded* Mechanics *to be of the greatest [significance] in the investigation of Natural things, and moderns, having rejected substantial forms and occult qualities, have undertaken to adapt the Phenomena of Nature to Mathematical laws, it seemed good to me to cultivate* Mathematics *in this Treatise, insofar as it relates to* Philosophy. *The Ancients really considered* Mechanics *twofold: as* Rational, *which proceeds accurately by Demonstrations, and* Practical. *All the manual arts belong to practical mechanics, from which the name mechanics was taken. But since artificers are not used to laboring with accuracy, it happens that* Mechanics *as a whole is distinguished from* Geometry *in this way: anything done accurately is referred to* Geometry; *anything done less accurately, to mechanics. The errors, however, are not of the art, but of the artificers. The one who works less accurately is a more imperfect mechanic, and the one who can work most accurately, this one would be the most perfect mechanic of all. For the descriptions of straight lines and circles, on which* Geometry *is founded, belongs to* Mechanics. Geometry *does not teach the drawing of these lines, but requires it. For it requires that the beginner learn to describe these [lines] accurately before he reaches the threshold of* Geometry; *then it teaches how problems may be solved through these operations. To describe straight lines and circles are problems, but not geometric ones. The solution of these is required from* Mechanics; *the use of the solutions is taught in* Geometry. *And* Geometry *prides itself that from so few principles brought from without, it accomplishes so many things. Therefore* Geometry *is founded on mechanical practice and is nothing else than that part of universal* Mechanics *that proposes and demonstrates the art of measuring accurately. But since the manual arts are chiefly employed in the moving of bodies, it happens that* Geometry *commonly refers to their magnitude, and* Mechanics *to their motion. In this sense* rational Mechanics *will be the science of motions that result from any forces whatso-*

ever, and also of the forces required to produce any motions, accurately proposed and demonstrated. This part of Mechanics *had been cultivated by the ancients among the* five powers *pertaining to manual arts, and they considered gravity (since it was not a manual power) hardly other than among the weights to be moved by those powers. But I consider philosophy rather than the arts, and I write concerning not manual, but natural powers, and consider chiefly those things that relate to gravity, levity, elastic force, the resistance of fluids, and similar forces, whether attractive or impulsive; and therefore I offer this work as the mathematical principles of philosophy. For the whole burden of philosophy seems to consist in this: from the phenomena of motions to investigate the forces of nature, and then from these forces to demonstrate the other phenomena; and to this end I directed the general propositions in the first and second books. In the third book I have set forth an example of this subject in the explication of the system of the world. For there, through propositions demonstrated mathematically in the first books, are derived from the celestial phenomena the forces of gravity by which bodies tend toward the sun and the individual planets. Then from these forces, through additional mathematical propositions, are deduced the motions of planets, comets, the moon, and the sea. I wish we could derive other phenomena of nature by the same kind of reasoning from mechanical principles. For I am induced to suspect for many reasons that they may all depend upon certain forces by which the particles of bodies, by some causes so far unknown, are either mutually impelled toward one another and cohere in regular figures, or are repelled and recede from one another. Since these forces are unknown, philosophers have up to now examined nature in vain. But I hope that the principles set down here will shed some light either on this method of philosophy or on some truer method.*

The most intelligent and universally erudite Mr. Edmund Halley *worked zealously in the publishing of this work. He assisted me not only in correcting the printing errors and preparing the geometrical figures, but he was also the source of my undertaking to publish them. For when he had received from me my demonstrations of the figure of the celestial orbits, he did not desist from urging me to communicate these to the* Royal Society, *who afterward, by their kind encouragement and requests, led me to think of publishing them. But after I had begun to consider the inequalities of the lunar motions, then I had also undertaken to attempt some other matters relating to the laws and measures of gravity and of other forces: [e.g.,] the figures that would be described by bodies attracted according to given laws; the motions of several bodies among themselves; the motions of bodies in resisting mediums; the forces, densities, and motions of mediums; the orbits of the comets, and other such matters. I thought that publication must be deferred to another time so that I might research these matters and might publish them together. What relates to the lunar motions (being incomplete) I have put all together in the corollaries of Proposition 66, to avoid being obliged to propose and distinctly demonstrate the several things contained there in a method more expansive than the subject deserved and interrupt the series of the other propositions. Some things, found out after the rest, I chose to insert in places less suitable, rather than to change the number of the propositions and the citations. I earnestly entreat that everything be read honestly, and that the defects, in matters so difficult, be not so much refuted, as investigated by the Readers' new efforts, and be kindly made complete.*

ON THIS WORK
OF MATHEMATICAL PHYSICS

OF THE MOST OUTSTANDING MAN
MOST LEARNED ISAAC NEWTON

an egregious distinction of our time and race.

Behold, for you, the pattern of the celestial globe,
the scales of divine mass, the computations of Jupiter,
which laws, while composing the beginnings of things,
the Creator did not wish to defile,
and fixed the foundations of the eternal work.
The inmost spaces of the conquered heaven lie open,
nor is the force obscured that rotates the farthest orbs.
The sun, residing on his throne, commands all things to tend toward him
by inclination and descent, nor does he allow the sidereal paths to be moved
through the vast void by a straight path,
but he attracts them to himself as center, in unmoved orbits.
Now lies open the veering path of the comet, once a source of dread,
nor are we now astonished at the phenomena of the bearded stars.
From this work we learn at last the reason that the silvery moon advances with steps unequal,
As if she scorned to suit her pace to numbers, to this day unknown to any astronomer;
Why the seasons return and the hours move forward.
We learn too with what forces roaming Cynthia impels the ebbing sea,
now, with waves subdued, it leaves the seaweed behind, and bares the
sands, esteemed by sailors; and now, in alternate turns, crashing on the highest shores.

Questions that tormented the minds of ancient scholars so often,
And even now trouble our scholars in vain with raucous strife,
We see clearly, learning routing the cloud of ignorance.
And now in no gloominess does error aggravate doubts for those
Whom the brilliance of sublime genius has allowed to penetrate
The homes of the gods and to scale the heights of heaven.

 Arise, Mortals, cast off earthly cares;
And from this [book] distinguish the powers of a heaven-born mind,
Removed, far and wide, from the life of the herds.
The person who through written tablets ordered the repression
Of murder, theft, and adultery, and crimes of perjured guile,
Or the one who gave to roving peoples cities with walls,
Was a founder; or the one who blessed the peoples with Ceres' gift
And who pressed from grapes the mitigation of cares;
Or the one who showed how to meld sounds inscribed on a reed of
 the Nile;
And thus to reveal voices to the eyes, did lighten the human lot:
Offsetting thus the miseries of life with some felicity.
But now we are admitted to the banquet of the gods; we are allowed
To deal with the laws of the celestial globe; and now
Those things enclosed and hidden in the blind earth lie revealed;
As are the unchanging order of things; and what past centuries
Of earth's long history have concealed.

 Now celebrate in songs with me,
You who delight in enjoying heavenly nectar,
NEWTON, who has set forth such things, unlocking the treasures of
 hidden truth,
NEWTON, dear to the muses,
So richly through his mind has Phoebus cast
The radiance of his own divinity.
Nearer to the gods no mortal may approach.

EDMUND HALLEY

ON THE MOTION OF BODIES

BOOK ONE

SECTION 1

*On the method of first and last ratios,
with the help of which the following [lemmas] are proved.*

Lemma 1.

Quantities, as well as ratios of quantities, that constantly tend to equality in a given time, and in that way are able to approach each other more closely than for any given difference, come ultimately to be equal.

If you deny it, let their ultimate difference be D. They are therefore unable to approach nearer to equality than for the given difference D; contrary to hypothesis.

Lemma 2.

If in any figure AacE *comprised of the straight lines* Aa *and* AE *and the curve* acE, *there should be inscribed any number of parallelograms* Ab, Bc, Cd, *etc.* [generated] *by equal bases* AB, BC, CD, *etc., and sides* Bb, Cc, Dd, *etc., parallel to the side* Aa *of the figure, and* [if] *the parallelograms* aKbl, bLcm, cMdn, *etc., should be completed, and then the width of these parallelograms should be diminished, and the number should be increased indefinitely:* [then] *I assert that the last ratios that the inscribed figure* AKbLcMdD, *the circumscribed* AalbmcndoE, *and the curvilinear* AabcdE, *are equal ratios.* [See fig. A.1.]

For the difference of the inscribed and circumscribed figures is the sum of the parallelograms $Kl + Lm + Mn + Do$, that is (because the bases of all are equal), the rectangle beneath the base Kb of one, and the sum Aa of their heights, and that is the rectangle $ABla$. But this rectangle, because

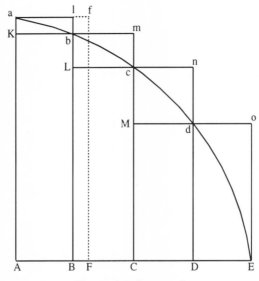

Figure A.1 Lemma 2.

its width *AB* is diminished indefinitely, comes to be less than any given [area]. Therefore, by Lemma 1, the inscribed and circumscribed figures, and even more by far the intermediary curvilinear figure come ultimately to be equal. Which was to be proven.

Lemma 3.

The same last ratios are also of equality when sides AB, BC, CD, *etc., of the parallelograms are unequal, and all are diminshed indefinitely.*

For let *AF* be equal to the maximum width, and let the parallelogram *FAaf* be completed. This will be greater than the difference of the inscribed and circumscribed figures, but with its width *AF* diminished indefinitely, it will come to be less than any given rectangle.

Corollary 1. Hence the ultimate sum of the vanishing rectangles coincides in every part with the curvilinear figure.

Corollary 2. And the rectilinear figure, which is comprised of the chords of the vanishing arcs *ab, bc, cd,* etc., ultimately coincides much more with the curvilinear figure.

Corollary 3. So too the rectilinear figure, which is circumscribed by the tangents of the same arcs.

Corollary 4. And consequently these ultimate figures (with respect to their perimeters *acE*) are not rectilinear, but the curvilinear limits of rectilinear [figures].

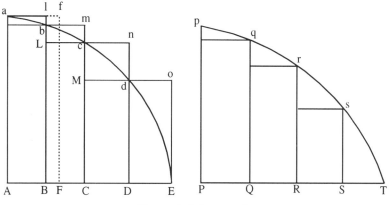

Figure A.2 Lemma 4.

Lemma 4.

If in two figures AacE and PprT there should be inscribed (as above) two series of parallelograms, and if the number of both should be the same, and when the widths are diminished indefinitely, the last ratios of parallelograms in one figure should be individually the same as the parallelograms in the other figure; [then] I assert that the two figures AacE and PprT are in the same ratio to one another. [See fig. A.2.]

For as the individual parallelograms are to each other, so (by compounding) does the total sum come to be one to the other, and so, too, figure to figure, the former figure, of course, being (by Lemma 3) to the former sum, and the latter figure to the latter sum in a ratio of equality.

Corollary. Hence if two quantities of any kind are divided in any way into the same number of parts, and, when their number is increased and their size diminished indefinitely, those parts maintain a given ratio to one another, the first to the first, the second to the second, and the rest to the rest in their sequence; [then] their totals will be to each other in that same given ratio. For if the parallelograms in the figures of this lemma should be assumed as parts of each other, [then] the sums of the parts will always be as the sums of the parallelograms; and consequently, when the number of parts and parallelograms is increased, and the size diminished indefinitely, [they will be] in the last ratio of parallelogram to parallelogram, that is (by hypothesis) in the last ratio of part to part.

Lemma 5.

[In] similar figures, all sides which correspond to each other mutually, curvilinear as well as rectilinear, are proportional, and their areas are in the doubled ratio of their sides.

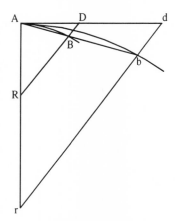

Figure A.3 Lemma 6.

Lemma 6.

If any arc AB given in position should be subtended by the chord AB, and [if] at some point A in the middle of its continuous curvature, it should be touched by the straight line AD extended in either direction; then [if] points A and B should approach each other and coalesce; I assert that the angle BAD [generated] by the chord and tangent, would be diminished indefinitely and would ultimately vanish. [See fig. A.3.]

For extend AB to b, and AD to d, and, with points A and B coalescing, with no part AB of Ab lying any longer within the curve, it is clear that this straight line Ab will either coincide with the tangent Ad, or will be drawn between the tangent and the curve. But the latter case is contrary to the nature of curvature, and therefore the prior case obtains. Which was to be proven.

Lemma 7.

With the same suppositions, I assert that the last ratio of the arc, chord, and tangent to each other is the ratio of equality. (See the figures of Lemmas 6 and 8.)
[See fig. A.4.]

For extend AB and AD to b and d, and parallel to the secant BD, draw bd. And let the arc Ab be similar to the arc AB. And, the points A and B coalescing, the angle dAb, by the lemma above, would vanish; and thus the straight lines Ab and Ad and the intervening arc Ab will coincide, and accordingly they will be equal. Hence the straight lines AB and AD, which are always proportional to these, and the intervening arc AB, will have the last ratio of equality. Which was to be proven.

Corollary 1. Hence if through B parallel to the tangent there should be

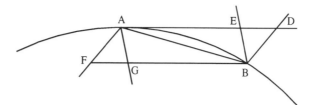

Figure A.4 Lemma 7.

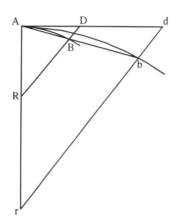

Figure A.5 Lemma 8.

drawn *BF* perpetually cutting at *F* any straight line *AF* passing through *A*, this will ultimately have the ratio of equality to the vanishing arc *AB*, since, when the parallelogram *AFBD* is completed, it always has the ratio of equality to *AD*.

Corollary 2. And if through *B* and *A* several straight lines *BE*, *BD*, *AF*, and *AG* are constructed, intersecting the tangent *AD* and its parallel *BF*, the last ratio of all the lines *AD*, *AE*, *BF*, *Bg*, and the chord and arc *AB* to each other will be the ratio of equality.

Corollary 3. And accordingly all these lines in any discussion about last ratios can be used mutually in place of each other.

Lemma 8.

If the given straight lines AR *and* BR *constitute with the arc* AB, *its chord* AB, *and the tangent* AD *the three triangles* ARB, ARB, *and* ARD; *and then [if] the points* A *and* B *approach each other; I assert that the ultimate forms of the vanishing triangles is one of similarity, and the last ratio, of equality.* [See fig. A.5.]

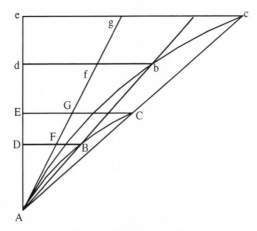

Figure A.6 Lemma 9.

For let *AB*, *AD*, and *AR* be extended to *b*, *d*, and *r*. Draw *rbd* parallel to *RD*, and construct arc *Ab* similar to arc *AB*. As the points *A* and *B* coalesce, the angle *bAd* vanishes, and accordingly the three triangles *rAb*, *rAb*, and *rAd* will coincide; and on that account they are similar and equal. And hence [triangles] *RAB*, *RAB*, and *RAD*, always similar and proportional to these [triangles], will come to be ultimately similar and equal to each other. Which was to be proven.

Corollary. And hence those triangles can mutually be used in place of each other in every proof of last ratios.

Lemma 9.

If the straight line AE *and the curve* AC *given in position mutually intersect at the given angle* A; *and* [*if*] BD *and* EC *should be applied as ordinates to that straight line at any given angle, meeting the curve at* B *and* C; *and* [*if*] *the points* B *and* C *approach point* A; [*then*] *I assert that the areas of triangles* ADB *and* AEC *will be ultimately to each other in the doubled ratio of the sides.* [See fig. A.6.]

For in *AD* extended take *Ad* and *Ae* proportional to *AD* and *AE*, and erect the ordinates *db* and *ec* parallel and proportional to the ordinates *DB* and *EC*. Extend *AC* to *c*, and construct the curve *Abc* similar to *ABC*, and let each curve be touched at *A* by the straight line *Ag*, and let the ordinates be cut by it at *F, G, f,* and *g*. Then let the points *B* and *C* coalesce with the point *A*, and with the angle *cAg* vanishing, the curvilinear areas *Abd* and *Ace* will coincide with the rectilinear [areas] *Afd* and *Age*; and thus, by Lemma 5, they will be in the doubled ratio of the sides *Ad* and *Ae*.

But always proportional to these areas are the areas *ABD* and *ACE*, and to these sides, the sides *AD* and *AE*. Therefore also the areas *ABD* and *ACE* are ultimately in the doubled ratio of the sides *AD* and *AE*. Which was to be proven.

Lemma 10.

Spaces which a body describes at the urging of any standard force are, at the very start of motion, in the doubled ratio of the times. [See fig. A.6.]

Let the times be expressed by the lines *AD* and *AE*, and the velocities generated by the ordinates *DB* and *EC*; and the first spaces described by these velocities will be as the first areas *ABD* and *ACE* described by these ordinates: that is, at the very beginning of motion (by Lemma 9) in the doubled ratio of the times *AD* and *AE*. As was to be proven.

Corollary 1. And hence it is easily concluded that in proportional times similar parts of similar figures of bodies describing deviations that are generated when equal forces are applied to the bodies similarly in those same parts, and that are measured by the positions of the figures to which the bodies would arrive in the same proportional times without those same forces, [these deviations] are as nearly as possible the squares of the times in which they are generated.

Corollary 2. But deviations that are generated by proportional forces similarly applied are as the forces and the squares of the times conjointly.

Lemma 11.

The vanishing subtense of an angle of contact is ultimately in the doubled ratio of the subtense of the conterminous arc. [See fig. A.7.]

Case 1. Let *AB* be the arc, *AD* its tangent, *BD* the subtense of the angle of contact perpendicular to the tangent, and *AB* the subtense of the arc. Perpendicular to the latter subtense *AB* and to the tangent *AD*, erect *AG* and *BG*, meeting at *G*; then let the points *D*, *B*, and *G* approach the points *d*, *b*, and *g*; and let *J* be the intersection of the lines *BG* and *AG*, ultimately occurring when the points *D* and *B* approach up to *A*. It is obvious that the distance *GJ* can be less than any assigned one.

Now (from the nature of circles passing through points *ABG* and *Abg*) AB^2 is equal to $AG \times BD$, and AB^2 is equal to $Ag \times bd$, and hence the ratio Ab^2 to Ab^2 is composed of the ratios *AG* to *Ag* and *BD* to *bd*. But since *JG* can be assumed less than any assigned length, it can be arranged that the ratio *AG* to *Ag* differs from the ratio of equality less than for any assigned difference, and thus as the ratio AB^2 to Ab^2 differs from the ratio *BD* to *bd* less than for any assigned difference. By Lemma 1 there is, therefore, the last ratio AB^2 to Ab^2 equal to the last ratio *BD* to *bd*. Which was to be proven.

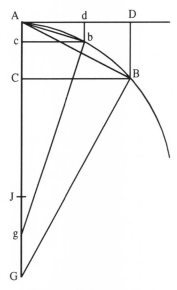

Figure A.7 Lemma 11.

Case 2. Now let *BD* be inclined to *AD* at any given angle, and there will always be the same last ratio as before and hence also the same AB^2 to Ab^2, as was to be proven.

Case 3. Moreover, although the angle *D* may not be given, the angles *D* and *d* will always verge toward equality and approach nearer to each other than for any assigned difference, and hence, by Lemma 1, will ultimately be equal, and accordingly, the lines *BD, bd* will be in the same ratio to each other as before. Which was to be proven.

Corollary 1. From this, since the tangents *AD* and *Ad*, the arcs *AB* and *Ab*, and their sines *BC* and *bc* are made ultimately equal to the chords *AB* and *ab*; their squares will also be ultimately as the subtenses *BD* and *bd*.

Corollary 2. The [area of the] rectilinear triangles *ADB* and *adb* are ultimately in the tripled ratio of the sides *AD* and *ad*, and in the three-halves ratio of the sides *DB* and *db*, since the definition [of area] is in the compounded ratio of the sides *AD* and *DB*, *Ad* and *db*. Thus also the triangles *ABC* and *Abc* are ultimately in the tripled ratio of the sides *BC* and *bc*.

Corollary 3. And since *DB* and *db* are ultimately parallel and in the doubled ratio of their [sides] *AD* and *AD*, the ultimate curvilinear areas *ADB* and *Adb* (from the nature of a parabola) will be two-thirds of the rectilinear triangles *ADB* and *Adb*, and the segments *AB* and *Ab* one-third of the same triangles. And hence these areas and these segments will be in the tripled ratio both of the tangents *AD* and *Ad* and of their chords and arcs *AB* and *Ab*.

Scholium

As for the rest, we suppose in all these things that the angle of contact is neither infinitely greater than the angles of contact which circles contain with their tangents nor infinitely less than these same [angles]; that is, the curvature at point A is neither infinitely small nor infinitely large, but the interval AJ is of finite magnitude. For DB can be taken proportional to AD^3, in which case no circle can be constructed through point A between the tangent AD and the curve AB, and hence the angle of contact will be infinitely less than the circular ones. And by a similar argument if DB is made successively proportional to AD^4, AD^5, AD^6, AD^7, etc., there will be a series of angles of contact continuing to infinity, any succeeding one of which is infinitely smaller than the preceding one. And if DB is made successively proportional to AD^2, $AD^{3/2}$, $AD^{4/3}$, $AD^{5/4}$, $AD^{6/5}$, $AD^{7/6}$, etc., there will be another series of infinite angles of contact, the first of which is of the same class as the circular ones, the second infinitely greater than it, and any succeeding one infinitely greater than the preceding one. But also between any two of these angles can be inserted a series of intermediate angles proceeding either way to infinity, of which any succeeding one will be infinitely greater than the preceding one. For example, between the terms AD^2 and AD^3 may be inserted the sequence $AD^{13/16}$, $AD^{11/5}$, $AD^{9/4}$, $AD^{7/3}$, $AD^{5/2}$, $AD^{8/3}$, $AD^{11/4}$, $AD^{14/5}$, $AD^{17/6}$, etc. And again there can be inserted between any pairs of angles in this sequence a new series of intermediate angles differing from one another by infinite intervals. And nature knows no limit.

What has been demonstrated about curved lines and the areas they contain may easily be applied to the curved surfaces and contents of solids. Indeed I have placed these lemmas first in order to avoid the tedium of deducing complicated proofs *ad absurdum* in the manner of ancient geometers. For proofs are rendered more compact by the method of indivisibles. But since the hypothesis of indivisibles is rather difficult, and moreover that method is considered less geometric, I have preferred to reduce proofs of the following matters to the last sums and ratios of vanishing quantities and the first ones of nascent quantities, that is, to the limits of these sums and ratios, and to place proofs of those limits first as concisely as I could. For the same is accomplished by these [proofs] as by the method of indivisibles, and when the principles have been proven, we shall employ them more safely. Hence in what follows, if I ever consider quantities as consisting of particles, or if for straight line segments I use minute curved ones, [then] I want it understood that these are not indivisibles, but vanishing divisibles, not the sums and ratios of definite parts, but the limits of such sums and ratios; and that the force of proofs must always be referred to the method of the preceding lemmas.

There is an objection, rather futile, that there is no ultimate proportion of vanishing quantities, since, before they have vanished, there is no last, and when they have vanished, there is none. But also by that same argument it can be contended that there is no ultimate velocity of a body proceeding to a certain position. For before the body reaches the position, there is no ultimate [velocity], and when it has reached it, there is no velocity. And the response is easy. For the ultimate velocity, is to be understood that with which the body moves, not before it reaches the last position and motion ceases, nor afterward, but at the moment when it reaches it, that is, that very velocity at which the body reaches the ultimate position and at which motion ceases. And likewise by the last ratio of vanishing quantities must be understood the ratio of quantities not before they vanish, nor afterward, but at which they vanish. Correspondingly, moreover, the first ratio of nascent quantities is the ratio at which they come into being. And their first and last sum is that at which they begin or cease to be (or to be increased or diminished). There exists a limit which their velocity can attain at the end of motion, but not exceed. This is the ultimate velocity. And there is an equal ratio of all quantities and proportions beginning and ceasing. Since this limit is certain and determined, the problem to determine it is truly a geometrical one. Indeed all things geometrical are legitimately employed in determining and demonstrating other geometrical matters.

It can also be argued that if the last ratios of vanishing quantities are given, the ultimate magnitudes will also be given; and thus every quantity will consist of indivisibles, contrary to what Euclid demonstrated on incommensurables in his tenth book of *Elements*. But really this objection is based on a false hypothesis. Those last ratios with which quantities disappear are not true ratios of ultimate quantities, but limits toward which ratios of quantities decreasing without limit are always approaching, and which they can follow more closely, for any given distance, but can never exceed, nor reach before the quantities diminish into infinity. The matter may be understood more clearly in infinitely large terms. If two quantities of a given distance are increased to infinity, the last ratio of these will be given, that is, the ratio of equality, and yet the ultimate or maximal quantities will not be given, that is, their ratio. Therefore in the following, if ever, intending an easy mental picture of things, I speak of the smallest quantities possible either as vanishing or as ultimate, [then] be careful not to understand quantities determined in magnitude, but consider them always diminished without end.

SECTION 2

Of the invention of centripetal forces.

Proposition 1. Theorem 1.

The areas that bodies driven in orbits describe with radii having been constructed to a stationary center of forces, lie in stationary planes and are proportional to the times. [See fig. A.8.]

Let the time be divided into equal parts, and in the first part of the time let a body by its innate force describe the straight line AB. The same body would then, if nothing impeded it, proceed directly to c in the second part of the time, describing the line Bc equal to the same AB (by Law 1), so that, when the radii AS, BS, and cS were drawn to the center, areas ASB and BSc would be made equal.

Now when the body comes to B, let the centripetal force act with one great impulse, and let it make the body deflect from the straight line Bc and proceed along the straight line BC. Parallel to the same BS, let cC be extended, meeting BC at C, and when the second interval of time is finished, the body (by Corollary 1 of the Laws) will be found at C, in the same plane with the triangle ASB. Join S and C and the triangle SBC which, because of the parallels SB and Cc, will be equal to the triangle SBc and hence also to the triangle SAB.

By a similar argument, if the centripetal force should act successively at C, D, E, etc., making the body in separate moments of time describe the separate straight lines CD, DE, EF, etc., then these will lie in the same plane, and the triangle SCD will be equal to the triangle SBC, SDE to SCD, SEF to SDE. In equal times, therefore, equal areas are described in a stationary plane; and in combining, whatever are the sums of areas $SADS$ and $SAFS$ among themselves, they are as the times of descriptions.

Now let the number of triangles be increased and their width be diminished indefinitely, and their ultimate perimeter ADF (by the fourth corollary of the third lemma) will be a curved line; hence the centripetal force by which a body is perpetually drawn away from the tangent of this curve, will act uninterruptedly; but whatever areas $SADS$ and $SAFS$ described which are always proportional to the times of the descriptions, will be proportional to the same times in this case. Which was to be proven.

Corollary 1. In nonresisting media, if the areas are not proportional to the times, [then] the forces are not directed along the path of the radii.

Corollary 2. In all media, if the description of the areas is accelerated, [then] the forces are not directed along the path of the radii, but in consequence deviate from there.

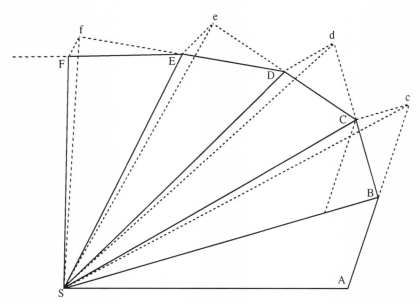

Figure A.8 Proposition 1.

Proposition 2. Theorem 2.

Every body that, when it is moved along some curved line with a radius having been constructed to a point [that is] either stationary or advancing uniformly in a rectilinear motion, describes areas proportional to the times around that point, is urged on by a centripetal force being directed to the same point.

Case 1. For, every body which is moved in a curved line is diverted from the rectilinear path by some force acting on it (by Law 1). Moreover, that force by which the body is diverted from the rectilinear path and is made to describe in equal times around a stationary point S the smallest equal triangles possible SAB, SBC, SCD, etc.; [such a force] acts at position B along a line parallel to cC (by *Elements,* I, 40, and Law 2): that is, along line BS; and at position C along a line parallel to dD: that is, along line CS, etc. It acts, therefore, always along lines being directed to that stationary point S. Which was to be proven.

Case 2. And by the fifth Corollary of the Laws, it is just the same whether the surface on which the body describes the curvilinear figure remains at rest, or whether the same surface is moved uniformly along a straight line, together with the body, the figure described, and its point S.

Scholium

A body can be urged by a centripetal force compounded of several forces. In this case the sense of the proposition is that the force which is compounded of all, is directed to the point S. Further, if any force acts along a line perpendicular to the described surface area, this will make the body deflect from the plane of its motion, but it will neither increase nor decrease the quantity of the described surface area, and, moreover, it must be neglected in the compounding of forces.

Proposition 3. Theorem 3.

Every body that, with a radius having been constructed to the center of another arbitrarily moving body, describes areas proportional to the times around that center, is urged by a force compounded of the centripetal force being directed toward the other body, and of the whole accelerative force by which the other body is urged.

For (by Corollary 6 of the Laws) if any body is urged along parallel lines by a new force which is made equal and opposite to that by which the other body is urged the first body will continue to describe the same areas around the other body as before; but the force by which the other body was being urged will now be canceled by the force equal and opposite to it, and moreover (by Law 1) that other body will either remain at rest or will be moved uniformly in a straight line, and the first body, at the urging of the difference of forces, will proceed to describe areas proportional to the times around the other body. The difference of forces, therefore (by Theorem 2), is directed to that other body as the center. Which was to be proven.

Corollary 1. Hence if one body, with a radius having been constructed to another body, describes areas proportional to the times, and [if] from the total force (whether single or compounded of several forces, according to the second corollary of the laws) by which the first body is urged [there] is subtracted the whole accelerating force by which the other body is urged (by the same corollary of the laws); [then] the whole remaining force by which the first body is urged will be directed toward the other body as center.

Corollary 2. And if those areas are as closely as possible proportional to the times, [then] the remaining force will be directed as closely as possible to the other body.

Corollary 3. And conversely, if the remaining force is directed as closely as possible to the other body, [then] those areas will be as closely as possible proportional to the times.

Corollary 4. If a body [A] with a radius having been constructed to another body [B] describes areas which, when compared with the times, are greatly unequal, and [if] that other body [B] either remains at rest or is moved uniformly in a straight line, [then] the action of centripetal force being directed to that other body [B] is either nothing or is mixed and compounded with the very powerful actions of other forces; and the entire force, compounded of all, if the forces are several, is directed to another center [C] (whether stationary or moving), around which there is an equal description of areas. The same obtains when the other body [B] moves with any motion whatever, if only there is assumed a centripetal force which remains after subtracting every force acting on that other body [B].

Scholium

Since the equable description of the areas is an indication of the center toward which is directed that force by which a body is most affected, while that body is kept in its orbit by the force being directed to the center; and [since] every circular motion is correctly said to be made around that center, by whose force the body is drawn away from the rectilinear motion and is kept in orbit; [then] why should we not use in what follows the equable description of areas as an indication of the center around which every circular motion [i.e., general curved motion] is performed in free space?

Proposition 4. Theorem 4.

For bodies that describe different circles with equable motion, the centripetal forces are directed toward the centers of those circles, and are to one another as the squares of the arcs described in the same time divided by the radii of the circles.
[See fig. A.9.]

Let the bodies B and b orbiting on the circumferences of circles BD and bd describe in the same time the arcs BD and bd. Since by their innate force alone they would describe the tangent lines BC and bc equal to these arcs, it is clear that the centripetal forces are those that perpetually draw bodies back from the tangents toward the circumferences of the circles, and hence these are to each other in the first ratio of the nascent intervals CD and cd; they tend, in fact, toward the centers of circles by Theorem 2 because the areas described by the radii are set proportional to the times.

Let figure tkb be similar to figure DCB, and then, by Lemma 5, the line segment CD will be to the line segment kt as the arc BD to the arc bt; and also, by Lemma 11, the nascent segment tk is to the nascent segment dc as

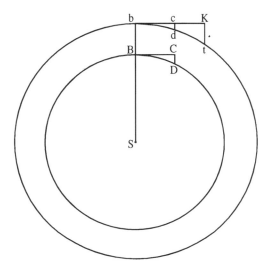

Figure A.9 Proposition 4.

bt^2 to bd^2, and, in like manner, the nascent line segment DC is to the nascent line segment dc as BD x bt to bd^2, or, what is the same thing, as BD x bt / Sb to bd^2 / Sb, and thus (from equal ratios bt / Sb and BD / SB) as BD^2 / SB to bd^2 / Sb. Which was to be proven.

Corollary 1. Hence the centripetal forces are as the squares of the velocities divided by the radii of the circles.

Corollary 2. And reciprocally as the squares of the periodic times divided by the radii, so are these forces to each other. That is (if I may speak as a geometer) these forces are in a ratio compounded of the doubled ratio of the velocities directly and in a simple ratio of the radii inversely; also in a ratio compounded of a simple ratio of the radii directly and of the squared ratio of the periodic times inversely.

Corollary 3. From this, if the periodic times are equal, both the centripetal forces and the velocities will be as the radii, and conversely.

Corollary 4. If the squares of the periodic times are as the radii, [then] the centripetal forces are equal, and the velocities in the halved ratio of the radii, and conversely.

Corollary 5. If the squares of the periodic times are as the squares of the radii, [then] the centripetal forces are reciprocally as the radii, and the velocities equal, and conversely.

Corollary 6. If the squares of the periodic times are as the cubes of the radii, [then] the centripetal forces are reciprocally as the squares of the radii; but the velocities are in the halved ratio of the radii, and conversely.

Corollary 7. [In the cases] in which bodies describe similar parts of any similar shapes whatsoever having centers similarly placed; all of these same [statements] about times, velocities, and forces follow from the demonstration of the preceding [corollaries] applied to these very cases.

Scholium

The case of the sixth corollary holds true in celestial bodies (as our compatriots Wren, Hooke, and Halley have also concluded independently), and for this reason I have decided to explain more fully in what follows those [questions] that concern centripetal force decreasing in the doubled ratio of the distances from the centers.

Further, with the help of the preceding demonstration, the proportion of a centripetal force to any known force, such as that of gravity, is also determined. For, in the time a body traverses the arc *BC*, that [centripetal] force, which at the very beginning of its motion is equal to the square of the arc *BD* divided by the diameter of the circle, impels [the body] through the space *CD*; [and since] every body with the same force continued always in the same region, would describe spaces in the doubled ratio of the times; [then], in the time a revolving body describes any given arc, that [centripetal] force will bring it about that the same body progressing in a straight line describes a space [*S*] equal to the square of that arc applied to the diameter of the circle; and thus [that centripetal force] is to the force of gravity as that space [*S*] is to the space which [the body] describes in falling heavily in the same time. And by propositions of this kind, Huygens in his outstanding tract on the oscillating clock, compared the force of gravity with centrifugal forces of revolving [bodies].

The preceding [statements] can also be demonstrated in this way. In any circle let a polygon of any number of sides be described. And if a body, in moving at a given velocity along the sides of the polygon, is reflected from the circle at each angle of the polygon, the force with which it impinges on the circle at each reflection will be as its velocity. And thus the sum of the forces in a given time will be as that velocity and number of reflections conjointly: that is, (if the species of the polygon is given) as the [product of the] length described in that given time and that length divided by the radius of the circle: that is, as the square of that length divided by the radius. Thus, if the polygon, with its sides diminished infinitely, coincides with the circle, then the sum of the forces in a given time will be as the square of the arc in a given time is divided by the radius. This is the force by which the body urges the circle, and equal to this is the opposite force by which the circle continuously repels the body toward the center.

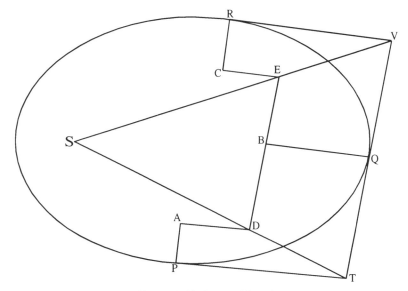

Figure A.10 Proposition 5.

Proposition 5. Problem 1.

Given, in any places, a velocity by which a body describes a given figure by forces directed to any common center, to find that center. [See fig. A.10.]

Let three straight lines *PT, TQV,* and *VR,* meeting at *T* and *V,* touch the figure described at the same number of points, *P, Q,* and *R.* Let *PA, QB,* and *RC* be erected perpendicular to the tangents, and inversely proportional to the velocities of the body at the points *P, Q,* and *R,* from which they are erected: that is, as *PA* is to *QB* so is the velocity at *Q* to the velocity at *P,* and *QB* is to *RC* as the velocity at *R* is to the velocity at *Q.* Through the ends of the perpendiculars, *A, B,* and *C,* let *AD, DBE,* and *EC* be constructed at right angles, meeting at *D* and *E.* And *TD* and *VE,* when drawn, will meet at the required center *S.*

For since the body at *P* and *Q,* with radii constructed to the center, describes areas proportional to the times, and since those areas described are at the same time as the velocities at *P* and *Q* constructed respectively on the perpendiculars dropped from the center to tangents *PT* and *QT,* those perpendiculars will be as the velocities reciprocally, and thus as the perpendiculars *AP* and *BQ* directly, that is, as perpendiculars dropped from point *D* to the tangents. Hence it is easily understood that points *S, D,* and *T* are in a straight line. And by a similar argument points *S, E,* and *V* are also in a straight line; and accordingly center *S* is found at the meeting of straight lines *TD* and *VE.* Which was to be proven.

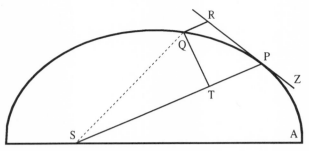

Figure A.11 Proposition 6.

Proposition 6. Theorem 5.

If a body P *by revolving around the center* S, *should describe any curved line* APQ, *and if the straight line* ZPR *should touch that curve at any point* P, *and if to this tangent from any other point* Q *of the curve,* QR *should be drawn parallel to the distance* SP, *and if* QT *should be dropped perpendicular to the distance* SP; [*then*] *I assert that the centripetal force would be reciprocally as the solid* SP² x QT² / QR, *provided that the quantity of that solid that ultimately occurs when the points* P *and* Q *coalesce is always taken.* [See fig. A.11.]

In the indefinitely small figure *QRPT* the nascent line segment *QR* is, given the time, as the centripetal force (by Law 2), and, given the force, as the square of the time (by Lemma 10); and hence, when neither is given, as the centripetal force and the square of the time conjointly; and hence the centripetal force is as the line segment *QR* directly, and the square of the time inversely. The time, however, is as the area *SPQ*, or its double *SP* x *QT*, that is, as *SP* and *QT* conjointly, and hence the centripetal force is as *QR* directly and *SP*² into *QT*² inversely, that is, as *SP*² x *QT*² / *QR* inversely. Which was to be proven.

Corollary. Hence if any figure is given, and on it a point toward which a centripetal force is directed, it is possible for the law of centripetal force to be found which will make a body orbit on the perimeter of that figure. Specifically the solid *SP*² x *QT*² / *QR* reciprocally proportional to this force must be computed. We shall give examples of this procedure in the following problems.

Proposition 7. Problem 2.

Let a body be orbited on the circumference of a circle; there is required the law of centripetal force being directed to some given point on the circumference. [See fig. A.12.]

Let *SQPA* be the circumference of a circle, *S* the center of centripetal force, *P* a body carried on the circumference, and *Q* a nearby position into

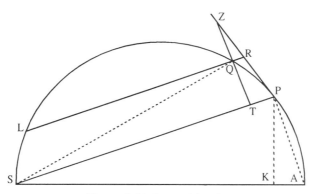

Figure A.12 Proposition 7.

which it will be moved. To the diameter SA and to the straight line SP drop the perpendiculars PK and QT, and through Q draw LR parallel to SP, reaching the circle at L and the tangent PR at R, and let TQ and PR come together at Z.

Because of the similarity of the triangles ZQR, ZTP, and SPA, there will be RP^2 (that is, $QR \times LR$) to QT^2 as SA^2 to SP^2. Therefore $(QR \times LR) \times (SP^2 / SA^2) = QT^2$. Multiply these equals by SP^2/QR, and, with points P and Q coalescing, let SP be written in place of LR.

Thus SP^5 / SA^2 will be made equal to $QT^2 \times SP^2 / QR$. Therefore the centripetal force is reciprocally as SP^5 / SA^2, that is (because SA^2 is given), as the fifth power of the distance SP (by the corollary of Theorem 5). Which was to be found.

Proposition 8. Problem 3.

Let a body be moved on a circle PQA: *for this effect there is required the law of centripetal force being directed to a point at such a distance that all lines* PS *and* RS *constructed to it can be considered as parallels.* [See fig. A.13.]

From the center of circle C let radius CA be drawn cutting those parallels perpendicularly at M and N, and let CP be joined. Because the triangles CPM and TPZ, that is, TPQ (by Lemma 8), are similar, there is CP^2 to PM^2 as PQ^2, that is, PR^2 (by Lemma 7) to QT^2, and from the nature of the circle, the rectangle $QR \times (RN + QN)$ is equal to the square of PR. But as the points P and Q coalesce, $RN + QN$ becomes equal to $2PM$. Therefore there is CP^2 to PM^2 as $QR \times 2PM$ to QT^2, and hence $QT^2 / QR = 2PM^3 / CP^2$ and so $QT^2 \times SP^2 / QR = 2PM^3 \times SP^2 / CP^2$. Therefore (by the corollary to Theorem 5) the centripetal force is reciprocally as $2PM^3 \times SP^2 / CP^2$, that is (after the determinate ratio $2SP^2 / CP^2$ is neglected), reciprocally as PM^3. Which was to be done.

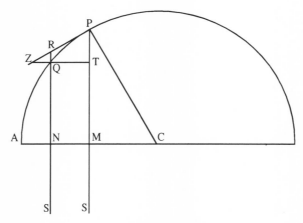

Figure A.13 Proposition 8.

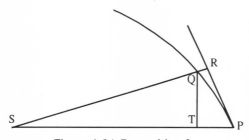

Figure A.14 Proposition 9.

Scholium

And by a similar argument a body will be moved on an ellipse, or also on a hyperbola or a parabola, by a centripetal force which is reciprocally as the cube of its ordinate being directed to a center of forces extremely far away.

Proposition 9. Problem 4.

Let a body be orbited on a spiral PQS intersecting all the radii SP, SQ, etc., at a given angle; there is required the law of centripetal force being directed to the center of the spiral. [See Fig. A.14.]

Let there be given the indefinitely small angle *PSQ*, and, because all the angles have been given, the figure *SPQRT* will be given in species. The ratio QT/RQ is therefore given, and there is QT^2/QR as QT, that is, as *SP*.

Now let the angle *PSQ* be changed in whatever manner, and the straight

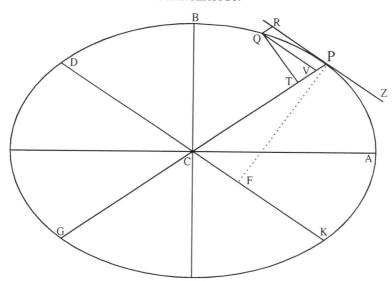

Figure A.15 Proposition 10.

line *QR* subtending the angle of contact *QPR* will be changed (by Lemma 11) in the doubled ratio of *PR* or *QT*. Therefore QT^2 / QR will remain the same as before, that is, as *SP*.

For this reason, $QT^2 \times SP^2 / QR$ is as SP^3, that is (by the corollary of Theorem 5), the centripetal force is [reciprocally] as the cube of the distance *SP*. Which was to be done.

Lemma 12.

All parallellograms described around a given ellipse are equal to each other. Understand the same for parallelograms described in a hyperbola around its diameters. This is evident from *Conics*.

Proposition 10. Problem 5.

Let a body be orbited on an ellipse; there is required the law of centripetal force being directed to the center of the ellipse. [See fig. A.15.]

Let *CA* and *CB* be the semi-axes of the ellipse; *GP* and *DK* conjugate diameters; *PF* and *QT* perpendiculars to these diameters; *QV* ordinate to the diameter *GP*;

and if parallelogram *QVRP* should be completed, there will be (from the *Conics*) $PV \times VG$ to QV^2 as PC^2 to CD^2, and (because of similar triangles QVT and PCF) QV^2 is to QT^2 as PC^2 to PF^2,

and on combining these ratios $PV \times VG$ to QT^2 as PC^2 to CD^2 and PC^2 to PF^2,

that is, VG to QT^2 / PV as PC^2 to $CD^2 \times PF^2 / PC^2$.

Write QR in place of PV, and (by Lemma 12) $BC \times CA$ in place of $CD \times PF$, and in addition (with the points P and Q coalescing) $2PC$ in place of VG, and, when the ends and middles are multiplied into each other, there will result $QT^2 \times PC^2 / QR = 2BC^2 \times CA^2 / PC$.

The centripetal force is therefore reciprocally as $2BC^2 \times CA^2 / PC$ (by the corollary to Theorem 5), that is (because $2BC^2 \times CA^2$ is given), as $1/PC$, that is, directly as the distance PC. Which was to be found.

Corollary 1. And therefore, conversely, if the force is as the distance, the body will be moved on an ellipse having its center at the center of forces, or perhaps on the circle into which the ellipse can degenerate.

Corollary 2. And there will be equal periodic times of revolutions made in all figures around the same center. For those times in the same ellipses are equal by Corollaries 3 and 7 of Proposition 4; but in ellipses having a common major axis, they are in turn as the total areas of the ellipses directly, and as parts of the areas described inversely in the same time; that is, as the minor axes directly, and the velocities of bodies on the principal vertices inversely, that is, as the axes directly to it, and the ordinates to the other axes inversely, and accordingly (because of the equality of the direct and inverse ratios) in the ratio of equality.

Scholium

If the ellipse, with the center going to infinity, should be turned into a parabola, a body will be moved on this parabola, and the force, now being directed to an infinitely distant center, will turn out equable. This is Galileo's theorem. And if the parabolic conic section, with the inclination of the plane to the conic section having been changed, should be turned into a hyperbola, the body will be moved on its perimeter, with the centripetal force changed into a centrifugal one.

SECTION 3

Of the motion of bodies in eccentric conic sections.

Proposition 11. Problem 6.

Let a body be revolved on an ellipse; there is required the law of centripetal force being directed to a focus of the ellipse. [See fig. A.16.]

Let S be a focus of the ellipse above. Let SP be drawn cutting not only the diameter of the ellipse DK at E, but also the ordinate QV in X, and let the parallelogram QXPR be completed.

It is clear that EP is equal to the semi-major axis AC, seeing that, when from the other focus H of the ellipse the line HI has been drawn parallel to CE (because CS and CH are equal), ES and EI would be equal, and hence EP would be half the sum of PS and PI, that is (because HI and PR are parallel, and the angles IPR and HPZ are equal), of PS and PH which are conjointly equal to the total axis 2AC.

Let the perpendicular QT be dropped to SP, and, after calling the principal *latus rectum* (or $2BC^2/AC$) of the ellipse L,

there will be L x QR to L x PV as QR to PV, that is, as PE (or AC) to PC;
and L x PV to GV x VP as L to GV;
and GV x VP to QV^2 as CP^2 to CD^2;
and (by Lemma 8) QV^2 to QX^2, is, with points P and Q coalescing, a ratio of equality;
and QX^2 or QV^2 is to QT^2 as EP^2 to PF^2, that is, as CA^2 to PF^2, or (by Lemma 12) as CD^2 to CB^2.

And when all these ratios are combined,

L x QR becomes to QT^2 as (AC to PC) x (L to GV) x (CP^2 to CD^2) x (CD^2 to CB^2),
that is, as AC x L (or $2CB^2$) x CP^2 becomes to PC x GV x CB^2,
that is, as 2PC to GV.

But with the points Q and P coalescing, 2PC and GV are equal. Therefore the proportionals of these, L x QR and QT^2, are also equal. Let these equals be multiplied by SP^2/QR, and L x SP^2 will be made equal to SP^2 x QT^2/QR.

Therefore (by the corollary of Theorem 5) the centripetal force is reciprocally as L x SP^2, that is, reciprocally in the doubled ratio of the distance SP. Which was to be found.

With the same brevity with which we transferred the fifth Problem to the parabola and the hyperbola, it would be possible to do the same here;

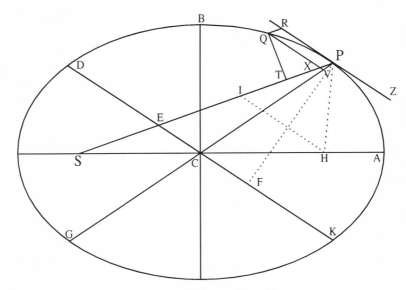

Figure A.16 Proposition 11.

but because of the dignity of the problem and its use in what follows, it will be no trouble to corroborate the other cases by [separate] proof.

Proposition 12. Problem 7.

Let a body be moved on a hyperbola; there is required the law of centripetal force being directed to a focus of the figure. [See fig. A.17.]

Let *CA* and *CB* be semi-axes of the hyperbola; *PG* and *KD* conjugate diameters; *PF* and *QT* perpendiculars to the diameters; and *QV* ordinate to the diameter *GP*. Let *SP* be drawn cutting both the diameter *DK* at *E* and the ordinate *QV* at *X*, and let the parallelogram *QRPX* be completed.

It is clear that *EP* is equal to the transverse semi-major axis *AC*, seeing that, when from the other focus *H* of the hyperbola the line *HI* has been drawn parallel to *CE* (because *CS* and *CH* are equal), *ES* and *EI* would be equal, and hence *EP* would be the half-difference of *PS* and *PI*, that is (because *HI* and *PR* are parallel, and the angles *IPR* and *HPZ* are equal), of *PS* and *PH*, whose difference is equal to the total axis 2*AC*.

Let the perpendicular *QT* be dropped to *SP*, and, after calling the principal *latus rectum* (or $2BC^2 / AC$) of the hyperbola *L*,

there will be *L* x *QR* to *L* x *PV* as *QR* to *PV*; that is, as *PE* (or *AC*) to *PC*; and *L* x *PV* to *GV* x *VP* as *L* to *GV*;

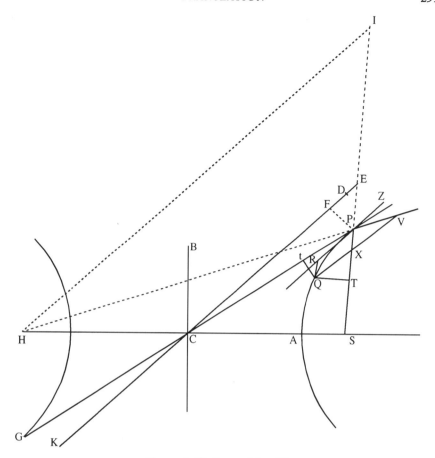

Figure A.17 Proposition 12.

and $GV \times VP$ to QV^2 as CP^2 to CD^2;

and (by Lemma 8) QV^2 to QX^2, becomes, with points P and Q coalescing, a ratio of equality;

and QX^2 or QV^2 is to QT^2 as EP^2 to PF^2, that is, as CA^2 to PF^2, or (by Lemma 12) as CD^2 to CB^2.

And when all these ratios are combined,

$L \times QR$ becomes to QT^2 as $(AC$ to $PC) \times (L$ to $GV) \times (CP^2$ to $CD^2) \times (CD^2$ to $CB^2)$,

that is, as $AC \times L$ (or $2\ CB^2$) $\times PC^2$ becomes to $PC \times GV \times CB^2$,

that is, as $2PC$ to GV.

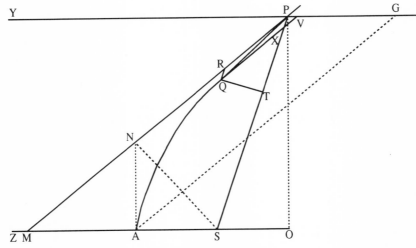

Figure A.18 Lemma 14.

But with points Q and P coalescing, $2PC$ and GV are equal. Therefore the proportionals to these, $L \times QR$ and QT^2, are also equal. Let these equals be multiplied by SP^2 / QR, and $L \times SP^2$ will be made equal to $SP^2 \times QT^2 / QR$.

Therefore (by the corollary of Theorem 5) the centripetal force is reciprocally as $L \times SP^2$, that is, reciprocally in the doubled ratio of the distance SP. Which was to be found.

In the same manner it is proven that a body, with this centripetal force changed to centrifugal, will be moved in a conjugate hyperbola.

Lemma 13.

The latus rectum of a parabola pertaining to any vertex is quadruple the distance of that vertex from the focus of the figure.
This is evident from the *Conics*.

Lemma 14.

A perpendicular which is dropped from the focus of a parabola to its tangent is a mean proportional between the distances of the focus from the point of contact and from the principal vertex of the figure. [See fig. A.18.]

For let APQ be the parabola, S its focus, A the principal vertex, P the point of contact, PO the ordinate to the principal diameter, PM a tangent meeting the principal diameter at M, and SN a perpendicular line from the focus to the tangent. Let AN be joined, and, because MS and SP, MN and NP, MA and AO are equals, straight lines AN and OP will be parallel,

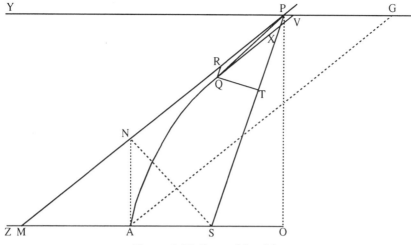

Figure A.19 Proposition 13.

and hence the triangle *SAN* will have a right angle at *A* and likewise for the equal triangles *SMN* and *SPN*. Therefore *PS* is to *SN* as *SN* to *SA*. Which was to be proven.

Corollary 1. PS² is to *SN²* as *PS* is to *SA*.

Corollary 2. And, because *SA* is given, *SN²* is as *PS*.

Corollary 3. And the meeting of any tangent *PM* with the straight line *SN*, which is perpendicular from the focus to it, falls on the straight line *AN*, which touches the parabola on the principal vertex.

Proposition 13. Problem 8.

Let a body be moved on the perimeter of a parabola; there is required the law of centripetal force being directed to the focus of this figure. [See fig. A.19.]

Let the construction of Lemma 14 remain, and let *P* be the body on the perimeter of the parabola, and then, from the position *Q* to which the body next moves, draw *QR* parallel to *SP* and *QT* perpendicular, and also *QV* parallel to the tangent and meeting both the diameter *YPG* at *V* and the distance *SP* at *X*.

Now because triangles *PXV* and *MSP* are similar, and the sides *SM* and *SP* of one are equal, the sides *PX* or *QR* and *PV* of the other are equal. But, from the *Conics*, the square of the ordinate *QV* is equal to the rectangle [generated] by the *latus rectum* and the segment *PV* of the diameter, that is (by Lemma 13), equal to the rectangle 4*PS* x *PV* or 4*PS* x *QR*; and, with points *P* and *Q* coalescing, the ratio *QV* to *QX* (by Lemma 8) is one of equality.

Therefore in this case QX^2 is equal to the rectangle $4PS \times QR$. But (because angles QXT, MPS, and PMO are equal), QX^2 is to QT^2 as PS^2 to SN^2, that is (by Corollary 1, Lemma 14), as PS to AS, that is, as $4PS \times QR$ to $4AS \times QR$, and therefore (by Proposition 9, Book 5 *Elements*), QT^2 and $4AS \times QR$ are equal.

Let these equals be multiplied by SP^2 / QR and $SP^2 \times QT^2 / QR$ will become equal to $SP^2 \times 4AS$; and therefore (by the corollary of Theorem 5) the centripetal force is reciprocally as $SP^2 \times 4AS$, that is, because $4AS$ is given, reciprocally in the doubled ratio of the distance SP. Which was to be found.

Corollary 1. From the last three propositions it follows that if any body P should depart from position P along any straight line PR, with any velocity, and is at the same time acted upon by a centripetal force that is reciprocally proportional to the square of the distance from the center, this body will be moved in one of the sections of conics having a focus at the center of forces; and conversely.

Corollary 2. And if the velocity with which the body departs from its position P should be such that the line element PR could be described in some minimal particle of time, and the centripetal force should be able to move the same body through the space QR in the same time; this body will be moved in some conic section whose *latus rectum* is that quantity QT^2 / QR, which would ultimately result when the line segments PR and QR are diminished infinitely. I place the circle [in the same category] with the ellipse in these corollaries, but I exclude the case where the body descends straight to the center.

Proposition 14. Theorem 6.

If several bodies should be revolved around a common center, and [if] the centripetal force should decrease in the doubled ratio of the distances from the center, [then] I say that the latera recta *of orbits are in the doubled ratio of the areas that bodies describe by radii constructed to the center in the same time.*

For (by Corollary 2, Problem 8) the *latus rectum L* is equal to the quantity QT^2 / QR, which would ultimately result when points P and Q coalesce. But the very small line QR, given the time, is as the generating centripetal force, that is (by hypothesis), reciprocally as SP^2. Therefore QT^2 / QR is as $QT^2 \times SP^2$; that is, the *latus rectum L* in the doubled ratio of the area $QT \times SP$. Which was to be proven.

Corollary. Hence the total area of the ellipse, and, proportional to it, the rectangle [generated] by the axes, is in the ratio compounded of the halved ratio of the *latus rectum* and the whole ratio of the periodic time.

Proposition 15. Theorem 7.

With the same suppositions, I say that the periodic times in ellipses are in the three-halved ratio of the transverse axes.

For the minor axis is the mean proportional between the major axis (which I call transverse) and the *latus rectum*, and therefore the rectangle [generated] by the axes is in a ratio compounded of the halved ratio of the *latus rectum* and the three-halved ratio of the transverse axis. But this rectangle, by the corollary of the sixth theorem, is in a ratio compounded of the halved ratio of the *latus rectum* and the whole ratio of the periodic time. Let the halved ratio of the *latus rectum* be taken away from each side and there will remain the three-halved ratio of the transverse axis equal to the ratio of the periodic time. Which was to be proven.

Corollary. The periodic times in ellipses, therefore, are the same as in circles whose diameters are equal to the major axes of ellipses.

Proposition 16. Theorem 8.

With the same suppositions, and with straight lines drawn to bodies that touch the orbits in the same places, and with perpendiculars dropped to these tangents from a common focus, I say that the velocities of the bodies are in a ratio compounded of the ratio of perpendiculars inversely, and the half ratio of the latera recta *directly.* [See fig. A.20.]

From the focus S to the tangent PR drop the perpendicular SY, and the velocity of the body P will be reciprocally in the halved ratio of the quantity SY^2 / L. For that velocity is as the smallest possible arc PQ described in a given particle of time, that is (by Lemma 7), as the tangent PR, that is (because PR is proportional to QT as SP to SY), as $(SP \times QT) / SY$, or as SY reciprocally and $SP \times QT$ directly; and there is $SP \times QT$ as the area described in a given time, that is (by Theorem 6), in the halved ratio of the *latus rectum*. Which was to be proven.

Corollary 1. The *latera recta* are in a ratio compounded of the doubled ratio of the perpendiculars and the doubled ratio of the velocities.

Corollary 2. The velocities of bodies at the greatest and smallest distances from the common focus are in a ratio compounded of the ratio of distances inversely and the halved ratio of the *latera recta* directly. For the perpendiculars are now those very distances.

Corollary 3. Thus the velocity in a conic section, at the smallest distance from the focus, is to the velocity in a circle at the same distance from the center, in the halved ratio of the *latus rectum* to that distance doubled.

Corollary 4. The velocities of bodies orbiting on ellipses at mean distances from the common focus are the same as those of bodies orbiting

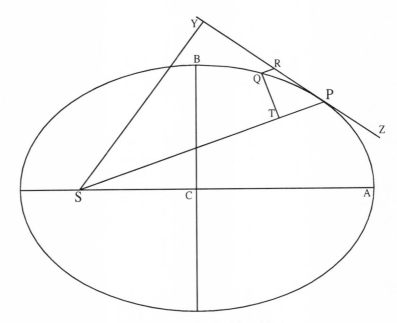

Figure A.20 Proposition 16.

on circles at the same distances, that is (by Corollary 6, Theorem 4), reciprocally in the halved ratio of the distances. For the perpendiculars are now [equal to] the semi-minor axes, and these are as the mean proportionals between the distances and the *latera recta*. Let this ratio be compound inversely with the halved ratio of the *latera recta* directly, and it will become the half ratio of the distances inversely.

Corollary 5. In the same [figure] or in equal figures, or even in unequal figures, whose *latera recta* are equal, the velocity of a body is reciprocally as the perpendicular dropped from the focus to the tangent.

Corollary 6. In a parabola the velocity is reciprocally in the halved ratio of the distance of the body from the focus of the figure; in the ellipse it is less than in this ratio, in the hyperbola greater. For (by Corollary 2, Lemma 14) the perpendicular dropped from the focus to the tangent of a parabola is in the halved ratio of the distance.

Corollary 7. In a parabola the velocity is everywhere to the velocity of a body orbiting on a circle at the same distance in the halved ratio of the number 2 to unity; in the ellipse it is less than in this ratio, in the hyperbola greater. For by the second corollary of this [proposition], the velocity on the vertex of a parabola is in this ratio, and by the sixth corollary of

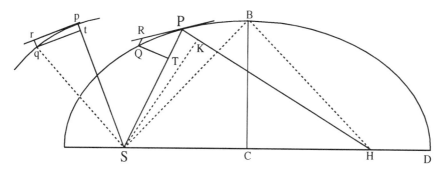

Figure A.21 Proposition 17.

this [proposition] and of the fourth theorem, the same proportion is preserved at all distances. Hence also in a parabola the velocity everywhere is equal to the velocity of a body revolving in a circle at half the distance; in the ellipse it is less, in the hyperbola greater.

Corollary 8. The velocity of [a body] orbiting on any conic section is to the velocity of [a body] orbiting on a circle at a distance of half of the *latus rectum* of the section, as that distance to the perpendicular dropped from the focus to the tangent of the section. It is evident by the fifth corollary.

Corollary 9. From this (by Corollary 6, Theorem 4) since the velocity of [a body] orbiting on this circle is to the velocity of [a body] orbiting in any other circle reciprocally in the halved ratio of the distances, similarly the velocity of [a body] orbiting on a conic section will become to the velocity of [a body] orbiting on a circle at the same distance, as the mean proportional between that common distance and half the *latus rectum* of the section to the perpendicular dropped from the common focus to the tangent of the section.

Proposition 17. Problem 9.

Supposing that the centripetal force be made reciprocally proportional to the square of the distance from its center, and that the absolute quantity of that force is known; there is required a line which a body will describe, when leaving from a given position with a given velocity along a given straight line. [See fig. A.21.]

Let the centripetal force being directed to point *S* be that which makes body *p* orbit in any given orbit *pq*, and let the velocity of this [body] at position *p* be known. Let the body *P* be released from the position *P* with a given velocity along line *PR*, and soon after let it deflect under the compulsion of the centripetal force into the conic section *PQ*. The straight line *PR*, therefore, will touch this at *P*.

In the same way let another straight line pr touch the orbit pq at p, and if from S perpendiculars are understood to be dropped to those tangents, there will be (by Corollary 1, [Proposition 16], Theorem 8) the *latus rectum* of the conic section to the *latus rectum* of the given orbit in a ratio compounded of the doubled ratio of the perpendiculars and the doubled ratio of the velocities, and therefore [the *latus rectum* of the conic section] is given. Let that be L.

In addition the focus S of the conic section is given. Let angle RPH be the complement of angle RPS to two right angles, and there will be given in position the line PH in which the other focus H is located. After the perpendicular SK has been dropped to PH, and the conjugate semi-axis BC has been erected, there is

$SP^2 - 2KP \times PH + PH^2$ (by Proposition 13, Book 2 of *Elements*)
$= SH^2$
$= 4CH^2$
$= 4BH^2 - 4BC^2$
$= (SP + PH)^2 - L \times (SP + PH)$
$= SP^2 + 2SP + PH + PH^2 - L \times (SP + PH).$

To each side let $2KP \times PH + L \times (SP + PH) - SP^2 - PH^2$ be added and there will come $L \times (SP + PH) = 2SP \times PH + 2KP \times PH,$
or, $SP + PH$ to PH as $2SP + 2KP$ to L.

Hence PH is given in length as well as position.

Specifically, if the velocity of the body at P should be such that the *latus rectum* L would be less than $(2SP + 2KP)$, [then] PH will lie on the same part of the tangent PR with the line PS; and thus the figure will be an ellipse, and it will be given both from the given foci S and H and from the [given] principal axis $(SP + PH)$.

But if the velocity of the body is so great that the *latus rectum* L is equal to $(2SP + 2KP)$, [then] the length of PH will be infinite, and as a result the figure will be a parabola having its axis SH parallel to the line PK, and from this it will be given.

But if the body is released from its position P with an even greater velocity, [then] the length PH must be taken on the other side of the tangent, and thus, with the tangent proceeding between the foci, the figure will be a hyperbola having its principal axis equal to the difference of the lines SP and PH, and from this it will be given. Which was to be found.

Corollary 1. Hence in every conic section, given the principal vertex D, the *latus rectum* L, and the focus S; the other focus H is given by taking DH to DS as the *latus rectum* is to the difference between the *latus rectum* and $4DS$. For the proportion $SP + PH$ to PH as $2SP$ to L, in the case of this corollary it would be $DS + DH$ to DH as $4DS$ to L, and, by dividing, DS to DH as $4DS - L$ to L.

Corollary 2. From there if the velocity of a body is given at its principal vertex D, the orbit will be found expeditiously, to be sure by taking its *latus rectum* to twice the distance DS in the doubled ratio of this given velocity to the velocity of the body revolving on a circle at the distance DS (by Corollary 3, Theorem 8), and then DH to DS as the *latus rectum* to the difference between the *latus rectum* and $4DS$.

Corollary 3. Hence even if a body should be moved on any conic section, and should be disturbed from its orbit by any impulse, the orbit can be ascertained in which it would afterward continue its course. For by combining the particular motion of the body with that motion which the impulse alone would generate, there will be obtained the motion with which the body, from the given place of the impulse, will depart from its position along a given straight line.

Corollary 4. And if that body, pressed by some outside force, is continually perturbed, its course will become known as nearly as possible by collecting the changes which that force induces at certain points and by estimating from an analogy of the sequence the continual changes at intermediate positions.

NOTES

CHAPTER 1. A SIMPLIFIED SOLUTION

1. Newton [prin. 1687] 1953, 401. All the translations from the 1687 edition of the *Principia* that are given in this book are by M. A. Rossi. See the Appendix of this book for a full translation of Sections 1, 2, and 3 of Book One of the *Principia*.

2. Newton [prin. 1687] 1953, Preface.

3. Cited in Westfall 1980, 154. The account comes from John Conduitt, who was Newton's assistant at the Mint in London and then the husband of Newton's niece, Clara Barton. See Westfall 1980, 154 n. 43 for a discussion of the various accounts of the "apple story." The particular tree was identified as the Flower of Kent that stood in the front garden of Newton's home in Woolsthorpe, and when it collapsed in the eighteenth century, a cutting was taken from it and grafted to another tree of the same type in the botanical gardens at Kew, just outside of London. Recently, another Flower of Kent was planted in the front garden of Newton's home in Woolsthorpe. So Newton's apple tree and the legend live on.

4. Bertoloni Meli 1993, 216.

5. In Blake's *The Marriage of Heaven and Hell* there is a figure holding what the editor identifies as "a huge pair of compasses, symbol of the Governor, or Reason, typified elsewhere by Newton with his compasses" (Blake [1789] 1975, Plate 5). A twelve-foot high bronze statue of Newton by Eduardo Paolozzi, based on Blake's portrait of Newton, is in the forecourt of the new British Library in London.

CHAPTER 2. AN OVERVIEW OF NEWTON'S DYNAMICS

1. For an extended discussion of this point, see Brackenridge 1982 and Brackenridge and Rossi 1979.

2. See Cohen 1983, 182–189, for a discussion of Descartes's early influence on Newton.

3. The gravitational force is given by M_g and the linear drag force by bV_T, where V_T is the terminal velocity and b is a constant. When the body reaches its

terminal velocity, the acceleration is zero, thus the net external force is zero, and thus $M_g = bV_T$. Therefore, both the terminal velocity V_T and the rate of fall are directly proportional to the mass M, as Aristotle stated.

4. "To solve the inverse problem, i.e., to ascertain under what law a force to a given center a given orbit can be described" (Lamb 1923, 249). That which was earlier called the *direct problem* (given the orbit, find the force) is now called the *inverse problem.*

5. Huygens first demonstrated the result in 1659, but it was not until 1703 that the demonstration was first published (posthumously). The result was published in 1673, however, but without demonstration.

6. Descartes [1644] 1983, 59.
7. Aristotle [c. 350 B.C.] 1961, 72.
8. Descartes [1644] 1983, 51.
9. Cited in Herivel 1965, 141.
10. Cited in Herivel 1965, 141.
11. For a discussion and resolution of this point, see Cohen, 1970, 159.
12. Descartes [1644] 1983, 141.
13. Osgood 1937, 102.
14. Bertoloni Meli 1993, 172. Also see Bertoloni Meli 1990 for an excellent essay that centers on the gulf between Newton's interpretation of centrifugal force in terms of the third law of motion and more modern interpretations.
15. Bertoloni Meli 1993, 41.
16. Bertoloni Meli 1993, 48.
17. "Definition 1 (in 1684) drew upon the lesson about circular motion that Hooke had taught him in 1679 to introduce a new word into the vocabulary of mechanics" (Westfall 1980, 411).
18. Westfall 1980, 383. Other scholars have taken up this claim. "What Hooke taught Newton was much more fundamental, namely, the correct way to analyze curvilinear motion" (Cohen 1985, 218).
19. Cited in Herivel 1965, 195.
20. For a discussion and documentation of Newton's reference to centrifugal endeavor following the 1687 *Principia,* see both Herivel 1965, 56–64, and Bertoloni Meli 1993, 182–183.
21. Herivel 1965, 42.
22. Koyré 1968, 95.
23. Newton [prin. 1726] [1729] 1968, Vol. 2: 195.
24. Newton [prin. 1726] [1729] 1968, Vol. 2: 197.
25. Dobbs 1991, 185–191.
26. For a discussion of Newton's continued interest in the dynamics of projectile motion, see Whiteside 1974, 6–8. In particular, note Newton's conjecture that the terrestrial path under (locally constant) gravitational force in a resistive medium would deviate from a parabolic path.
27. Dobbs 1991, 186.
28. Dobbs argues that such a mathematical solution is not sufficient. "Having broken with mechanical causation for gravity, however, Newton renewed his search for its cause. So far was he from accepting the nineteenth and twentieth centuries' argument that the mathematical laws of gravity suffice that Newton spent the rest of his life searching for a new causal explanation" (Dobbs 1991, 252). Dobbs fur-

ther argues that Newton's rejection of a mechanical explanation of gravitation provided him with the opportunity to consider "the possibility that gravitational phenomena might carry evidence of God's direct and immediate relationship with the cosmos" (Dobbs 1991, 210). She concludes, "And so it was to be for the next three decades that Newton gave preferential consideration to a spiritual cause for gravity" (Dobbs 1991, 190).

29. Whiteside 1964, 131.

30. Whiteside 1964, 135.

31. This distinction has no meaning for uniform circular motion, because for such motion the radius from the center sweeps out both equal angles and equal areas in a given time. There is no indication that this relationship in this special case played any role in Newton's development of the law for the general case of central forces.

32. Whiteside 1964, 136.

33. For a discussion of Newton's early knowledge of Kepler's area law in particular and of Newton's early astronomical texts in general, see Whiteside 1974, Vol. 6: 6 n. 15. In an earlier article Whiteside concludes that it is possible but unlikely that Newton knew of the area law as a student (Whiteside 1964, 124 n. 24).

34. Thoren 1974, 244.

35. For a discussion and documentation of the influence of Galileo's work on Newton's dynamics, see Herivel 1965, 35–41, and Cohen 1983, 132–133.

36. "That Newton's primary indoctrination in the principles of terrestrial and celestial motion came initially through his study of the published work of Descartes—rather than, as once thought, from reading Galileo and Kepler—is now firmly documented" (Whiteside 1974, Vol. 6: 3). For a full discussion and documentation of the influence of Descartes on Newton's dynamics, see Herivel 1965, 44–53.

37. Whiteside 1974, Vol. 6: 33 n. 12. Whiteside continues the discussion of Newton's presupposition of the sufficiency of the parabolic approximation and of the polygonal approximation and finds it wanting under modern analysis. It can only be used correctly, he claims, for infinitesimal arcs, but he concludes that "such sophisticated complexities do not there bedevil the logical cogency of its present proof" (Whiteside 1974, Vol. 6: 33 n. 19). Erlichson has taken issue with what he calls "the second order fallacy" as put forward by Whiteside and he claims that in "the limit . . . all three of the orbit models [polygonal, parabolic, or circular] converge to the same limit, the actual orbital arc" (Erlichson 1993, 254). My feeling is that the argument concerning how valid Newton's approximation is in light of modern standards has little importance in evaluating Newton's achievement. If anything, his ability to adapt the contemporary seventeenth-century kinematics of Galileo to the solution of the Kepler problem simply reveals his genius. Cohen discusses this topic in terms of Newton's transition from the earlier impulsive forces of collisions to the later continuous forces (Cohen 1970, 168).

38. Newton considers the ratio $QT^2 \times SP^2 / QR$, which has the dimensions of a solid but is *inversely* proportional to the force, rather than $QR / QT^2 \times SP^2$, which is *directly* proportional to the force. I have elected to use the direct proportion when not quoting directly from Newton's text.

39. In the revisions following the publication of the first edition, there is clear evidence of Newton's use of fluxions (both successful and unsuccessful) in dynamic

problems. But there is no evidence that he attempted to employ fluxional analysis in the preparation of the first edition, nor is there need of it, as his own geometric/limiting technique works quite well. See Whiteside 1970, 119.

40. The properties of an ellipse are: (1) an "evident" relationship that Newton demonstrates in the opening lines of Proposition 11; (2) a relationship developed from Proposition 31, Book 7, of Apollonius's *Conics*; and (3) a relationship developed from Proposition 15, Book 1, of Apollonius's *Conics*.

41. Two of the geometric properties are from similar triangles and one from the construction of a parallelogram.

42. Newton employs the Apollonian definition of the constant *latus rectum* L of the ellipse (i.e., $L : 2BC :: 2BC : 2AC$ or $L = 2\,BC^2 / AC$).

43. The circular approximation, in which the unique circle of curvature is used, is on a much firmer mathematical foundation than the parabolic approximation. Questions have been raised concerning the rigor of the latter but not of the former. See the detailed discussion in chapter 10.

44. Some critics have faulted Newton for apparently failing to include the element of time in his statement of the second law (i.e., the change in motion Δmv is given as proportional to the motive force rather than the *time rate* of change in motion $\Delta mv / \Delta t$). This position is refuted by Cohen in an excellent article in which he discusses the second law in terms of impulse $I = \Delta mv$ (Cohen 1970). In any event, note that before Newton states the second law he has, in Definition 8, anticipated the time dependence: "The motive quantity of a centripetal force [which Newton says in the next line he will call 'for brevity's sake by the name of Motive force'] . . . proportional to the motion it generates in *a given time*." Newton may be at fault in not stressing in the second law what he has already made clear in the earlier definition (i.e., that the change in motion takes place in *a given time*), but he is not confused, nor does he switch around. Newton does not need to make explicit use of the given time in any of the direct problems he solves in Sections 2 and 3 of Book One. When he needs it for the inverse problem in Proposition 39 of Book One, however, he uses it in a compact modern form: $F = m\Delta v / \Delta t$. "Force will be as the increment I of the velocity (i.e., Δv) directly and the time inversely (i.e., Δt)" (Newton 1687, 123).

45. Westfall 1980, 413.

46. See Cohen 1970 and 1983 for a full discussion of Newton's developing concept of force as it relates to impulsive and particulate time.

47. Bertoloni Meli 1993, 182.

48. Barbour 1989, 610.

CHAPTER 3. NEWTON'S EARLY DYNAMICS

1. "Like Galileo, Newton had the fortune to make a beginning in a tradition that had undergone continuous and careful development over a period of two thousand years. Genius he truly was, but, as he would be the first to acknowledge, he also stood 'on the shoulders of giants.' It was his successful assimilation of what had been done before, no less than what he discovered on his own, that enabled him to inaugurate the new direction in science now associated with his name" (Wallace 1988, 36).

2. See Westfall for an informative discussion of Newton's life as a student (Westfall 1980, 66–104). See Whiteside's introduction for a detailed discussion of Newton's mathematical education (Whiteside 1967, 1–15).

3. At that time in England the new year began legally on 25 March (old style) but the common practice was to begin it on 1 January (new style). Some people adhered to the legal practice, others used 1 January, and still others used double years (e.g., 1664/65) for the period between 1 January and 25 March. The calendar problem was even more complex than the date of the new year. The Julian calendar reform of 45 B.C. inserted a leap year every four years. The result was a great improvement but it still resulted in about three additional days in four centuries. By the later half of the sixteenth century, the date of the vernal equinox was about ten days earlier than it was in the first half of the fourth century, when the rules for the observance of Easter had been set. Therefore, in 1582, Pope Gregory XIII introduced yet another calendar reform in which ten days were omitted from that year and it was arranged to omit three leap years in the next four centuries (1700, 1800, and 1900 are thus leap-leap years and 2000 is a leap-leap-leap year). England did not adopt the Gregorian calendar until the middle of the eighteenth century, perhaps seeing it as popish interference. Thus, before 1700 England was ten days ahead of continental calendars and after 1700, which it observed as a leap year, it was eleven days ahead.

4. Herivel 1965, 133–135.

5. Cited in Herivel 1965, 153.

6. Cited in Herivel 1965, 129–130. The comparatively sophisticated material on circular motion quoted here appears undated in folio 1. The first date that appears in the *Waste Book* is "Jan. 20th 1664" (old style) and occurs in the margin of folio 10, which is concerned with simple collisions. Herivel argues that despite its location, the material in folio 1 was written after that date, 20 January 1664/65 (old style/new style).

7. In an early discussion of the text, Herivel stated, "Noteworthy is the fact that up to this point [the writing of this particular text] he had not obtained a formula giving the dependence of centrifugal force on the speed and the radius of the circle" (Herivel 1965, 10). When Herival introduced the text he stated, "There can be no doubt that the first two dynamical entries on this folio [this text is the first entry] have been made later than the discussion of circular motion beginning at Ax.-Prop. 20" (Herivel 1965, 129).

8. At this point in the text, Newton had first written and then canceled the following: "As if b is moved with one degree of motion through bn in one second of an hower then its force from the center n being continually (like the force of gravity) impressed upon a body during one second it will generate one degree of motion in that body." This statement, which confuses as much as it clarifies, appears in the manuscript in heavy square brackets, which are often used by Newton instead of cross-strokes to cancel material (Whiteside, personal correspondence). Herivel reproduces it as a parenthetical expression (Herivel 1965, 129).

9. In one sense, this statement can be seen in contemporary terms as the definition of an angle in radians. The angle θ in radians between two given radii is given by $\Delta S/R$, where ΔS is the arc length and R is the radius. The angle between the velocity vectors normal to the given radii is the same angle θ, and thus $\theta =$

$\Delta S / R = \Delta V / R$, where ΔV is the change in velocity that Newton describes as the "motion in a body begat by the force of endeavor from the center" and V is the velocity that Newton describes as the "force of the body's motion."

10. Cited in Herivel 1965, 146–147.

11. "If . . . three bodys move from the point a uniformly and in equal times, the first to d, the second to e, the third to c: then is the third's motion compounded of the motion of the first and the second" (Hall and Hall 1962, 15). Also see Newton's mathematical papers for other early notes on this topic (Newton [math. 1664–1666] 1967, Vol. 1: 390, 392).

12. Recall that the body moved from point a to point b between collisions in a time ΔT with the given orbital motion mv: the "force of motion." If there had not been a collision at point b, then in the next time interval ΔT, the ball would have gone to the point y, where the distance $by = v \Delta T = (mv)(\Delta T / m)$. If the ball had been at rest at point b and subjected to the impulse, then in the same time interval ΔT the ball would have gone to the point d (as demonstrated by the parallelogram rule), where the distance $bd = \Delta v \Delta T = (\Delta mv)(\Delta T / m)$. When these two expressions are solved for Δmv and mv in terms of the lines bd and by, then the ratio of the "force or pressure" (the impulse due to the collision at point b that generates the "change in the motion": Δmv) and the "force of motion" (the initial impulse that generates the orbital motion: mv) is determined. Thus, the ratio of "force or pressure" to the "force of motion" = $(\Delta mv) / (mv) = (bd)(m / \Delta T) / (by)(m / \Delta T) = (bd) / (by) = 2af / ab$, where $bd = 2fa$ and $by = ab$. It was demonstrated that $(2fa) / (ab) = (ab) / (fa)$, thus the "force or pressure" to the "force of motion" = ab / fa, as required. (Note that the deviation $yc = bd$ is a measure of the force in the sense that yc [or bd] is proportional to Δmv [and hence is proportional to the force] for a given $\Delta T / m$.)

13. The same result can be demonstrated using modern concepts. First, the "impulse" I is defined as the product of the force F and the time of collision δt, and second, the impulse is demonstrated to be equal to the change in momentum $m\Delta v$, where m is a given mass and Δv is the change in the velocity. Thus, the impulse associated with the collision at b is given by $I_b = m\Delta v$. The deviation yc is equal to the change in velocity Δv multiplied by the time between collisions, ΔT, or $yc = (\Delta v)(\Delta T) = (I_b / m)(\Delta T)$. The initial impulse I_o that is required to establish the orbital motion is equal to mv. The distance ab is the product of the velocity v and the time between collisions is ΔT, or $ab = (v)(\Delta T) = (I_o / m)(\Delta T)$. Thus, the ratio $yc / ab = (I_b / m)(\Delta T) / (I_o / m)(\Delta T) = I_b / I_o = 2fa / ab$, which is the result given by Newton above for the ratio of the "force or pressure" and the "force of b's motion."

14. It is also possible to reverse the procedure and use the definition of impulse to obtain the nature of the force from Newton's result. The $\Sigma(I) / I_o = \Sigma(F \delta t) / mv = F\Sigma(\delta t) / mv = FT / mv = F(2\pi r / v) / mv = 2\pi$. Solving for the force, one has $F = mv^2 / r$. Note that the nature of the force itself is locked into the result until the modern definition of the impulse employed: $\Sigma(I) = \Sigma(F \delta t) = F\Sigma(\delta t) = FT$, where T is the period ($T = 2\pi r / v$) and the sum is taken as the number of collisions increases without limit.

15. See the discussion of Proposition 4 in chapter 7 and the full text in the Appendix.

16. The editor of Newton's mathematical papers, D. T. Whiteside, sees it as evident that Newton assumes "that the infinitesimal arc is approximated to sufficient accuracy by a parabola whose diameter passes through the force-centre, with its deviation from the inertial tangent-line according proportional to the square of the time" (Whiteside 1974, Vol. 6: 33 n. 12).

17. Cited in Herivel 1965, 193–195. See Newton [math. 1664–1666] 1967, Vol. 1: 297–303. Hall first published and translated the manuscript (Hall 1957, 62–71). Herivel also published and translated it (Herivel 1965, 193–197) as well as Turnbull. The text itself is undated, but Hall, Herivel, and Turnbull all date it sometime before 1669. Mary Ann Rossi translates the Latin text given here.

18. Yoder 1988, 19–22. Huygens's original worksheet, dated "21 Oct. 1659," is reproduced on page 20.

19. Newton [math. 1664–1666] 1967, Vol. 1: 456.

20. Newton [math. 1664–1666] 1967, Vol. 1: 456. Since making the transcription of Newton's original English manuscript, the editor D. T. Whiteside has decided that the word given first as "will" is in fact "may" (personal correspondence). This topic will be discussed in further detail in chapter 8 when the alternate solutions to the problems are considered.

21. Newton [math. 1664–1666] 1967, Vol. 1: 252–255.

22. For an informative discussion of a very early numerical solution that is based upon curvature, see Nauenberg 1994a. "An iterative computational method to evaluate orbits for central forces is described, which is based on Newton's mathematical development of the concept of curvature started in 1664. This method accounts very well for the orbit obtained by Newton for a constant central force, and it gives convergent results even for forces which diverge at the center, which are discussed correctly in Newton's letter *without* using Kepler's area law" (Nauenberg 1994a, 221). This excellent article provides a view of Newton's early dynamics that bears upon the letter and diagram that he sent to Hooke in 1679. Nauenberg concludes that "the letter and diagram reveal that Newton's understanding of orbital dynamics at the time of his correspondence with Hooke was deeper than has been previously realized, in spite of the fact that he had not yet understood the significance of Kepler's law of areas" (Nauenberg 1994a, 243). See chapter 10 in this book for a discussion of the analytical solution that Nauenberg suggests Newton may have developed before he demonstrated Kepler's law of areas in 1679.

CHAPTER 4. THE PARADIGM CONSTRUCTED

1. Newton [corres. 1676–1687] 1960, Vol. 2: 297, letter 235.
2. Hooke [1679] 1945, Vol. 8: 27–28.
3. Newton [corres. 1676–1687] 1960, Vol. 2: 300, letter 236.
4. Cited in Cohen 1971, 293.
5. For a discussion of this "lost solution of 1679" and its relationship to the two solutions that Newton describes (one sent to Locke in 1690 and the one that appears in *On Motion*), see Brackenridge 1993. Also see the discussion of the "Locke solution" in chapter 8.
6. Cook 1991, 132–133. Cook speculates that Halley was in the vicinity of Cambridge to settle various matters of family business following the death (murder or

suicide) of his father in March of 1683 (old style). Halley's attention to family affairs immediately following his father's death might explain why he did not pursue Wren's challenge immediately.

7. Cited in Cohen 1971, 297.

8. The solution to the direct problem also provided a solution to the inverse problem if the uniqueness of the direct solution could be demonstrated. However, neither the solution he sent to Halley in 1684 nor the solution published in the 1687 edition of the *Principia* addressed that question directly. Newton did demonstrate in Problem 4 of the 1684 tract (which became Proposition 17 in the 1687 *Principia*) that if the initial orbit was a conic, then for all other dynamic conditions it would continue to be a conic. He took up the subject of the uniqueness of the solution more directly in the revised (1713) edition of the *Principia*. This topic will be discussed in more detail in subsequent chapters.

9. There remains then the nagging question of how much Newton's solution of the direct Kepler problem is indebted to the suggestion made in Hooke's initial correspondence of November 1679 concerning the elements of dynamics. There is no question that the correspondence eventually provided the stimulus that sent Newton back to consider the problem of planetary motion. Further, there is no question that in his early analysis of curvilinear motion, Newton was concerned with the Cartesian outward endeavor. In fact, in his second response to Hooke in 13 December 1679, Newton still referred to the outward centrifugal force and the inward gravitational force interacting when he wrote of the "*vis centrifuga* and gravity alternately overbalancing one another." Newton's next letter to Hooke came almost a year later, and it did not contain any comments upon his solution to the planetary problem. The solution that Newton produced for Halley in 1684, however, did not make mention of the Cartesian outward endeavor, but employed instead the newly coined term *centripetal* force in contrast to the *centrifugal* force. In this respect, the solution echoes Hooke's suggestion in which the centrifugal force does not play a necessary role. The two elements Newton employed in his solution, however, are the parabolic approximation from his pre-1669 analysis of uniform circular motion and the law of equal areas in equal times from his post-1679 demonstration. In chapter 2, I suggested that the close correspondence of the area law with observed planetary motion argued against celestial vortices, and that it was this demonstration that caused Newton to reject Cartesian celestial dynamics. From that perspective, Newton owes a debt to Hooke for the clarification, but not for the detailed method. Comparison of the post-1679 solution of the problem of elliptical motion with the pre-1669 solution of the problem of uniform circular motion demonstrates that the parabolic approximation is applied to both problems in a similar fashion. In the limit of small times, the force is assumed to be approximately constant, and thus inversely proportional to the square of the time and directly proportional to the radial displacement. The area law permits the method of the parabolic approximation to be extended to non-uniform motion, but the procedure is the same in both solutions. Hooke served to catalyze and clarify Newton's thoughts on celestial dynamics, but it was Newton who created the method of analysis.

10. The motion described in Problem 2 is, in fact, that which is now called simple harmonic motion. No reference is made to such an application of this type of

force. In chapter 10 of this book I will give a criterion for the case of solution based on the concept of curvature.

11. See Dobbs 1991, 185–191, for a detailed discussion of this point.

12. For a discussion of the various versions of the tract *On Motion,* see Whiteside 1974, 31 nn. 1–2. Also see Herivel 1965, 102–108, and Cohen 1971, 54–62. The version of the tract employed in this chapter is that found in the *Register Book* of the Royal Society (Vol. 6: 218–234) dated 10 December 1684.

13. See Whiteside 1974, Vol. 6: 31 n. 4, for a discussion of the terms *centripetal* and *centrifugal.*

14. See, for example, the discussion of the corollary to Proposition 4 in chapter 3 in which the term centrifugal is omitted from the 1687 edition but inserted into the 1713 edition of the *Principia.*

15. See Whiteside 1974, Vol. 6: 32 n. 9, for a discussion of Newton's early statement of this hypothesis.

16. As important as this relationship is to Newton's analysis, it came only as "a late marginal addition" (Whiteside 1974, Vol. 6: 33 n. 2). Perhaps the addition as an afterthought is a demonstration of how fundamental and "obvious" the relationship was to Newton's analysis.

17. Whiteside speculates that Newton may have been unaware that this lemma appears in Apollonius, and that his reference to the *Conics* may have been "an oblique reference to the ingenious proof by area-dissection given by Grégoire de Saint-Vincent" (Whiteside 1974, Vol. 6: 34 n. 14).

18. See Whiteside 1974, Vol. 6: 31 n. 2, for a discussion of the various versions of this text. For a full translation of *On the Motion of Spherical Bodies in Fluids,* see Hall 1962, 243–267.

19. In the 1713 edition of the *Principia,* Newton adds a series of new corollaries to the Proposition 1. The first three of these new corollaries provides at last Newton's explicit defense of his dynamic application of the parallelogram rule (see chapter 9).

20. Questions concerning the rigor of this limiting process have been of much interest to scholars. Whiteside claims that "Newton's proof of this fundamental generalization of Kepler's area law is more subtle and considerably less cogent than it may at first appear" (Whiteside 1974, Vol. 6: 35 n. 19). Newton makes no further defense.

21. The manuscript shows an insertion and deletion of the phrase "of the speeds or," indicating that Newton did consider expressing the statement in those terms. See Whiteside 1974, Vol. 6: 37 n. 20.

22. See Whiteside 1974, Vol. 6: 38–39 nn. 22–23, for a discussion of the infinitesimal involute arcs *CD* and *cd.*

23. Proposition 36, Book 3, Euclid [c. 300 B.C.] 1952, 64–66.

24. Kepler [1619] 1952, 1009–1085. See Whiteside 1974, Vol. 6: 38 n. 26, for a detailed discussion of Newton's concern with Kepler's third law.

CHAPTER 5. THE PARADIGM APPLIED

1. The solution to Problem 2 states that the force is directly proportional to the distance *SP*, which is a force that every modern textbook recognizes as a "simple

harmonic force." Nothing indicates here or in the *Principia* that this example is seen as anything other than a preliminary mathematical exercise to prepare the reader for Problem 3.

2. It is possible to apply Euclid's Proposition 32, Book 3, to this problem, and the presence of the perpendicular *PK* in Newton's drawing indicates that he may have done so (Euclid [c. 300 B.C.] 1952, 60–61). I find it simpler to employ the diameter *OP*.

3. When Newton attempted in 1684 to reconstruct for Halley the "lost solution of 1679" he mistakenly drew the conjugate diameters *DK* and *PG* as perpendiculars. He was unable to reproduce the solution until he noted the error. See chapter 8 for a discussion of the controversy concerning this lost solution. For an extended discussion of this diagram and the errors that appear in subsequent versions of it, see Brackenridge 1985.

4. Whiteside 1974, Vol. 6: 47 n. 45.
5. Apollonius [c. 200 B.C.] 1952, 787–788.
6. Dampier 1966, 152.

CHAPTER 6. THE PARADIGM EXTENDED

1. See Whiteside 1974, Vol. 6: 51 n. 62, for a discussion of this point.

2. "Newton is seemingly unaware that the technique here cited . . . is Halley's straightforward (if unacknowledged) borrowing from Kepler's *Astronomia Nova*" (Whiteside 1974, Vol. 6: 52 n. 63).

3. See Whiteside 1974, Vol. 6: 52 n. 65, for a demonstration and a reference to Apollonius.

4. Whiteside 1974, Vol. 6: 53 n. 66.

5. "This circuitous appeal to an auxiliary circle orbit is not, we may remark, at all necessary" (Whiteside 1974, Vol. 6: 56 n. 73). Whiteside then provides an alternate demonstration.

6. Bertoloni Meli 1993, 214.

7. In response to Euler's claim for this solution, one contemporary critic writes, "Even Euler—the overwhelmingly great Euler—reports on the *Principia* without having made certain of what he reports on!" (Weinstock 1989, 848). Weinstock strongly, if not modestly, disagrees with Euler's claim that this result is a solution to the inverse problem. Another scholar responds, however, that "this argument is based upon a profound delusion . . . (and) there is no doubt about the uniqueness" (Arnol'd 1990, 31–33). This point will be discussed in more detail in chapter 10.

8. Apollonius [c. 200 B.C.] 1952, 787–788.
9. Euclid [c. 300 B.C.] 1952, 39–40.

10. "The still more fundamental corollary that none but conic orbits are traversible in an inverse-square force—since for every initial speed v and angle of projection α a unique trajectory . . . may correspondingly be defined, so exhausting all possibilities of motion . . . is here taken by Newton to be self-evident" (Whiteside 1974, Vol. 6: 56 n. 73).

11. Whiteside 1974, Vol. 6: 58 n. 79. Notes 79 and 84 provide a full discussion of the details.

12. See Whiteside 1972, Vol. 5: 524 n. 1, for a discussion of this question and for a detailed example of such a calculation.
13. Whiteside 1974, Vol. 6: 58. See notes 79 and 84 for an extended detailed discussion of this method.
14. Newton [corres. 1688–1694] 1961, Vol. 3: 385, letter 461.
15. For an interesting discussion of Newton's calculation of the mass, surface gravity, and density of Jupiter, Saturn, and the earth relative to the sun in Proposition 8 of Book Three, see Garisto 1991, 42–48.
16. See Cohen 1971, 47–142, for a detailed chronology of writing and publication of the first edition.
17. Newton [corres. 1676–1687] 1960, Vol. 2: 480–481.

CHAPTER 7. THE *PRINCIPIA* AND ITS RELATIONSHIP TO *ON MOTION*

1. Newton [prin. 1687] 1953, 401.
2. For an extended discussion of this topic, see Cohen 1970.
3. The interested reader is referred to the detailed discussion of this scholium in chapter 11 of Barbour's excellent book (Barbour 1989, 598–644).
4. Plato [c. 350 B.C.] 1965, 51.
5. Aristotle [c. 350 B.C.] 1961, 80.
6. This view of space led Aristotle to reject the possibility of the existence of a "void." In fact, he argues that if a void did exist, then "a body would either continue in its state of rest or would necessarily continue in its motion indefinitely, unless interfered with by a stronger force" (Aristotle [c. 350 B.C.] 1961, 72). But since such an ideal state is never observed, Aristotle rejects the existence of a void. This statement by Aristotle of "Newton's first law of motion" is rarely noted in physics textbooks.
7. Galileo is then faced by the counterargument that he cannot therefore demonstrate that the earth is in motion. His response is to theorize that the tides result from the sloshing of the ocean due to the motion of the earth, much as water sloshes around in the bottom of a moving barge.
8. Descartes's relativism may well have its roots in the Inquisition's condemnation of Galileo's argument for the movement of the earth. Descartes was very much aware of that action and it has been suggested that this "scholium was Newton's response to Descartes's squirming before the Inquisition" (Barbour 1989, 598).
9. For a discussion of electromagnetic forces and relativity, see Brackenridge and Rosenberg 1970, 322–334.
10. Barbour 1989, 629.
11. For an extended discussion of Law 2 in terms of continuous and impulsive forces, see Cohen 1970, 178–185.
12. See Whiteside 1974, Vol. 6: 107 n. 39.
13. See the introduction to the translation in the Appendix for a further discussion of this point.
14. See Erlichson 1992a, 369–375, for an argument concerning the importance of this inverse theorem.

15. Whiteside 1974, Vol. 6: 199 n. 21.
16. Cohen 1971, 135.
17. Whiteside 1974, 137 n. 99. This error is continued without comment in all published editions of the *Principia*, and Whiteside appears to be the first person to point out the error. See also Whiteside 1974, 137 nn. 95–98, for a detailed discussion of this proposition.
18. In the 1687 *Principia*, the line QR is drawn parallel to the line SQ instead of parallel to the line SP. The demonstration is valid in the limit in either case but in every other example QR is drawn parallel to SP (i.e., the deviation QR is in the direction of the force at point P rather than at point Q). In the revised editions Newton changes the direction of QR in this proposition to conform to that practice.
19. The radius from the pole of the spiral to a general point is given by $r = r_o e^{(\text{ctn}\alpha)\theta}$, where α is the constant angle between the polar radius r and the tangent to the curve and θ is the polar angle.
20. The polar radius r is related to the radius of curvature ρ as follows: $r = \rho \, \text{ctn}\alpha$, where α is the constant angle between the radius and the tangent.
21. See Erlichson 1992b, 403–406, for a discussion of this proposition.
22. Specifically, the pole distance $SP = PC \sin(SPR)$, where the ratio of the sides is independent of the particular value of the radius of curvature PC. Consider for example a radius PC of 10 and a constant angle SPR of 30 degrees: then the pole distance SP is 5 and thus all the other sides, and hence all the ratios of the sides, are uniquely determined from the fixed angles.
23. The radius of curvature $\rho = r[1 + (r'/r)^2]^{3/2} / [1 - (r''/r) + 2(r'/r)^2]$ where $r' = dr/d\theta$ and $r'' = d^2r/d\theta^2$ and $r = Ce^{(\text{ctn}\alpha)\theta}$ is the polar equation for the equiangular spiral in terms of the polar radius r, the polar angle θ, and the constant angle α. Substitution of r, r', and r'' into the formula for the radius of curvature gives $\rho = r/\sin\alpha$ or $SP = PC \sin\alpha$. See Whiteside 1969, Vol. 3: 171 n. 330, for a discussion of Newton's development of the general expression for the radius of curvature in terms of polar coordinates.
24. None of the properties of the spiral are explicitly given by Newton in the text of the *Principia*. It is clear from his early work, however, that he was in command of such relationships and had obtained the general measure of curvature in both Cartesian and polar coordinates. See Whiteside 1974, Vol. 3: 171 n. 330. In a work published in 1730, just after Newton's death, John Clarke avoided the need for the curvature relationships by producing a "two point" solution for the spiral/pole problem (Clarke 1730, 164). For a discussion of the relative merits of the two solutions, see Erlichson 1992b, 403–406.
25. For a full discussion of this proposition, see Whiteside 1974, Vol. 6: 34 n. 14.
26. See the discussion of Proposition 1 and the introduction to the Appendix for possible implications of such a change.
27. This reference is to Book 1, Proposition 49, of the *Conics* (Apollonius [c. 200 B.C.] 1952, 661–663).
28. Whiteside discusses the development of this lemma from Newton's initial flawed proof to the final corrected one that appears here as Lemma 14 (Whiteside 1974, Vol. 6: 145 n. 119).
29. See Whiteside 1974, Vol. 6: 146 n. 124, for an extended discussion of the question. Additional references to more recent critics will be given in chapter 10.

30. See the conclusion to the discussion of Problem 4 in chapter 6 for the problem's role in the solution to the inverse problem.

CHAPTER 8. NEWTON'S UNPUBLISHED PROPOSED REVISIONS

1. Whiteside 1974, Vol. 6: 568 n. 1.
2. Whiteside 1974, Vol. 6: 568 n. 1.
3. Newton [corres. 1688–1694] 1961, Vol. 3: 384.
4. Newton [math. 1684–1691] 1974, Vol. 6: 573. *Proposed Proposition 6*. Where bodies describe all similar parts of similar figures in proportional times, their centripetal forces tend to centers similarly positioned in those figures and are to one another in a ratio compounded of the ratio of the heights directly and the doubled ratio of the times inversely.
5. Newton [math. 1684–1691] 1974, Vol. 6: 575–577. *Proposed Proposition 7*. If in two orbits proportional ordinates stand at any given angles on proportional abscissas, and the centers of forces are similarly located in the abscissas, bodies shall describe corresponding parts of the orbits in proportional times and the centripetal forces will be as the heights of the bodies directly and the squares of the times inversely. . . .

Corollary. Therefore if one of the orbits APB be a circle and the other orbit AQB any ellipse, and the point S be the center of both, since the force whereby a body revolves with uniform motion in a circle is given . . . the force whereby another body might simultaneously revolve in the ellipse will be as the height SQ of that body.

6. Newton [math. 1684–1691] 1974, Vol. 6: 579–581. *Proposed Proposition 8*. The force whereby any body P can revolve in any orbit APB whatever round the center S of force is to the force whereby another body P can revolve in the same orbit and in the same periodic time round any other center R of force as the product of the height of the first body and the square of the height of the second body, $SP \times RP^2$, to the cube of the straight line PT which the straight line ST parallel to the orbit's tangent cuts off from the height of the second body in the direction of that body.

Corollary 1. . . .

Corollary 2. . . . the centripetal force will be reciprocally as the square of the height PR.

Corollary 3. . . . will also in this case—where, that is, the ellipse has passed into a parabola—be reciprocally as the square of the height.

Corollary 4. . . . the force whereby a body can revolve in a hyperbola about its focus as center will be reciprocally as the square of the height.

7. Newton [math. 1684–1691] 1974, Vol. 6: 581. *Proposed Proposition 9*. If a body should, in a nonresisting space, revolve round a stationary center in any orbit whatever and describe any just barely nascent arc in a minimal time, and an "arrow" of the arc be drawn to bisect its chord and pass, when produced, through the center of force, then the centripetal force at the mid-point of the arc will be as that sagitta directly and the square of the time inversely.

Corollary 1. . . . the centripetal force will then be reciprocally as the "solid" $SP^2 \times QT^2 / QR$ . . .

Corollary 2. ...
Corollary 3. ... the centripetal force will be reciprocally as the "solid" $SY^2 \times P$.
Corollary 4. ...
Corollary 5. ...

8. Newton [math. 1684–1691] 1974, Vol. 6: 583. *Proposed Lemma 12.* If in any diameter *PG* of a conic there be taken, on its concave side, *PM* equal to the *latus rectum* pertaining to that diameter, and through the points *P* and *M* a circle be described to touch the conic at *P*, then this circle will have the same curvature as that conic at *P*.

9. Newton [math. 1684–1691] 1974, Vol. 6: 585–589. *Proposed Proposition 10.* Let a body move in the perimeter of the conic *PQ*: there is required the centripetal force tending to any given point *S*.

Corollary 1. If the force tends toward the center of the conic ...
Corollary 2. If the force tends toward the conic's focus ...
Corollary 3. If the force tends toward an infinitely distant point ...
Corollary 4. If the conic passes into a circle and the centripetal force should tend to a point given in its circumference ...

10. Whiteside 1974, Vol. 6: 581 n. 29.

11. Newton [math. 1664–1666] 1967, Vol. 1: 456. In Whiteside's original transcription of Newton's English manuscript, the word given here as *may* appeared as *will*. The change in meaning is quite dramatic: *may* indicates that Newton is giving one possible method for solving the problem while *will* indicates that something specific must follow. The editor, D. T. Whiteside, has informed me that he now believes that the word he first transcribed as "will" is in fact "may." In Herivel's transcription the word is given in brackets with a question mark: [will?] (Herivel 1965, 130). Clearly, Herivel also had difficulties in deciphering Newton's English script. Herivel, however, mistakenly transcribes Newton's "y^n" as "that" rather than as "then" (Herivel 1965, 130), which further obscure's Newton's intent and makes more difficult the choice of "will" or "may" by context (i.e., "*then* the force ... *may* be found" versus "*that* the force ... *will* be found").

12. Newton [math. 1684–1691] 1974, Vol. 6: 578.

13. *PT* is equal to *CA*, where *CA* is half the major axis of the ellipse. Newton had demonstrated this relationship in the proof to Proposition 11 in the 1687 edition and intended it to appear in an introductory lemma in the unpublished radical revision. See Whiteside 1974, Vol. 6: 580 n. 25.

14. Newton [math. 1684–1691] 1974, Vol. 6: 585.

15. For the central ellipse, Proposition 10, the line *PE* = *PS* = *PC* and, thus, the force is directly as *PC*. For the focal ellipse, Proposition 11, the line *PE* is a constant of the ellipse (equal to the semi-major axis *AC*) and thus the force is inversely proportional to the square of the distance *SP*. The remaining two published problems concerning circular paths are both dispatched in a similarly simple fashion as Corollaries 3 and 4 of this new Problem 2.

16. Whiteside 1974, Vol. 6: 589 n. 48.

17. Hall and Hall 1962, 293.

18. Ball 1893, 116.

19. Two copies of Newton's tract, both in English, survive in manuscript: one, at Cambridge, is an autograph and is not dated; the other, at Oxford, is a copy in the hand of Locke's amanuensis Brownover and is dated "March 1689," which is

presumably March 1690 (new style). The major portion of the manuscript is found in Ball 1893, 116–120. A complete copy appears in Hall and Hall 1962, 293–301, and in Herivel 1965, 246–254.

20. Between 1961 and 1969, four related articles appeared in *Archives internationales d'histoire des sciences* that gave rise to a debate which has continued intermittently ever since: Herivel 1961 and 1963, Hall and Hall 1963, and Westfall 1969. The discussion in this section of chapter 8 is taken from Brackenridge 1993. See Erlichson 1993 for an alternate argument concerning the "missing solution of 1679." For a suggestion that the curvature method may have been employed by Newton before 1679, see Nauenberg 1994 and my discussion in chapter 10 to follow.

21. Whiteside 1989, xv.

22. This special case is present in the copy of the manuscript at Cambridge but missing from the copy of the manuscript at Oxford.

23. Herivel 1965, 249.

24. It will be demonstrated in chapter 10 that the circular ratio $1 / (YS^2 \times PV) = 1 / (SP^2 \times \rho \times \sin^3\alpha)$ where SP is the radius, ρ the radius of curvature, and α the sine of the angle between the tangent and the radius. The radius of curvature is equal at aphelion and perihelion and, for any point on the major axis of the ellipse, the angle α is the same. Thus, the force is as the inverse square of the distance SP to any point on the major axis. This relationship was called to my attention by D. T. Whiteside.

CHAPTER 9. NEWTON'S PUBLISHED RECAST REVISIONS

1. Newton [math. 1661–1675] 1959, Vol. 1: 245.

2. Whiteside 1959, Vol. 6: 245 n. 76.

3. Whiteside 1959, Vol. 1: 156 n. 288. The radius of curvature DC is given in terms of z, where z is dy/dx. (Newton [math. 1664–1666] 1967, Vol. 1: 157.)

4. Newton [math. 1664–1666] 1967, Vol. 1: 159.

5. Newton [math. 1664–1666] 1967, Vol. 1: 169.

6. Whiteside 1974, Vol. 3: 171 n. 330. The Cartesian expression
$\rho = [1 + (dy/dx)^2]^{3/2} / [d^2y/dx^2]$
can be written in terms of a parameter t as $[x'^2 + y'^2]^{3/2} / [x'y'' - y'x'']$,
where $x = x(t)$, $y = y(t)$, and $x' = dx/dt$, $x'' = d^2x/dt^2$, and so on.
It may also be expressed in terms of polar coordinates as follows:
$[r^2 + r'^2]^{3/2} / [r^2 - rr'' + 2r'^2]$ or $r[1 + (r'/r)^2]^{3/2} / [1 - (r''/r) + 2(r'/r)^2]$
where $r' = dr/d\theta$ and $r'' = d^2r/d\theta^2$.
If $z = r'/r$, then $z' = (r''/r) - z^2$,
and thus the radius of curvature is given by $r[1 + z^2]^{3/2} / [1 + z^2 - z']$,
which is the form given in note 330, cited above, with $r = y$ and $z' = z$.

7. Lemma 11 is also employed in the solution of the direct spiral/pole problem of Proposition 9, both in the first and revised published editions.

8. Newton employs here the term "versed sine" or "versine" for the line AC (and in Theorem 5 for the line QR). The lines AC and QR approach the versine only in the limit as the two points on the curve come together. (The versine = $1 - \text{cosine}$.)

9. Whiteside 1974, Vol. 6: 117 n. 54.

10. "The purpose of the new opening Corollaries 1–3 is to clarify the way in which the basic parallelogram of 'forces' serves to yield a measure of an impressed force in terms of the velocity increment BV (= cC) which its instantaneous impulse at B toward the centre S produces in the time that the orbiting body so impelled would otherwise move uniformly from B to c at a pristine inertial speed represented by $AB = Bc$, thereby producing a new resultant uniform motion at B, now towards C, which is correspondingly represented by $AV = BC$" (Whiteside 1974, Vol. 6: 542 n. 13).

11. Whiteside 1974, Vol. 6: 543 n. 13.

12. "Newton here introduces a happy generalization of what, in the elementary trigonometry of the circle, was in his day a familiar technical term [sagitta] for the versine—one which in Arabic is at least as old as . . . early ninth century, and in the medieval Latin West was already used . . . about 1330" (Whiteside 1974, Vol. 6: 545 n. 17).

13. "In a preliminary recasting . . . Newton toyed with the notion of making this derived property basic as a variant 'Prop. VI. theor. V'" (Whiteside 1974, Vol. 6: 548 n. 19).

14. See Brackenridge 1988, 474, for a reproduction of the original manuscript page.

15. Cited in Brackenridge 1988, 473.

16. In modern vector notation, F_O is the component of the force F_S perpendicular to the tangent (i.e., $F_O = F_S \cos\theta$, where $\cos\theta = PV / PX$). See Brackenridge 1988, 466, or Brackenridge 1990, 14, for a discussion of this point.

17. See Whiteside 1974, Vol. 6: 548 n. 25, and Brackenridge 1988, 473–475.

18. The lines SY and OP are parallel because both are normal to the tangent YP. Thus, the angle $YSP = OPS$. Further, since the line $OP = OV$, then the angle $OPS = OSP$. Thus, since the angle $VPA = 90°$, the triangles SYP and VPA are similar.

19. Both SY and PF are normal to the set of parallel lines PR and DK and the point S is identical to the center C.

20. For a full solution of this alternate method, see Brackenridge 1988, 470–472.

21. The line YS does appear in the first edition, but it is used only in reference to Proposition 16 (i.e., the same figure is used for Propositions 10, 11, and 16).

22. Whiteside 1974, Vol. 6: 600 n. 1.

CHAPTER 10. NEWTON'S DYNAMICS IN MODERN MATHEMATICAL DRESS

1. Such a force is called a central force, and it can be argued simply that it gives rise to motion restricted to the plane defined by the initial position and velocity vectors.

2. See any standard textbook on mechanics, such as Symon 1971, 93–94.

3. The polar orbital equation often appears as $d^2(r^{-1}) / d\theta^2 = [-r^{-1} - (m / L^2) r^{-2} F_r]$, where the expectation is that the force F_r will be given and the path $r = r(\theta)$ will be found. See, for example, Symon 1971, 109.

4. Whiteside 1969, Vol. 3: 171 n. 330.

5. By definition $z = \text{ctn}(\alpha)$.
Thus, $(1 + z^2)^{3/2} = (1 + \text{ctn}(\alpha)^2)^{3/2} = (1 / \sin(\alpha)^2)^{3/2} = \sin^{-3}(\alpha)$.
Also, $z = (1/r)dr/d\theta = -r\, d(r^{-1})/d\theta$.
Thus, $z' = dz/d\theta = d[-r(d\, r^{-1})/d\theta]/d\theta$
$= -(dr/d\theta)(d(r^{-1})/d\theta - r(d^2(r^{-1})/d\theta^2)$
$= r^2[(d(r^{-1})/d\theta)^2 - r\, d^2(r^{-1})/d\theta^2]$.
Therefore,
$(1 + z^2) - z' = (1 + r^2(d(r^{-1})/d\theta)^2) - (r^2(d(r^{-1})/d\theta)^2 - r\, d^2(r^{-1})/d\theta^2)$
$= (r)\ (r^{-1} + d^2(r^{-1})/d\theta^2)$,
as was required.

6. Newton [math. 1664–1666] 1967, Vol. 1: 456. "If the body b moved in an Ellipsis then its force in each point (if its motion in that point bee given) may bee found by a tangent circle of Equall crookednesse with that point of the Ellipsis."

7. Whiteside 1974, Vol. 6: 42 n. 30.

8. Brackenridge 1992, 254.

9. Newton [math. 1664–1666] 1967, Vol. 1: 245–297.

10. Newton [math. 1664–1666] 1967, Vol. 3: 169.

11. On 13 December 1679, Newton sent Hooke a letter on orbital dynamics containing a diagram for the orbit of a body subject to a constant central force. In the letter Newton did not reveal any details of his method of construction beyond a cryptic reference to "the method of indivisibles." This figure has been the subject of considerable scholarly concern, with most finding it to be wanting in many respects. Nauenberg, however, argues that Newton's method was correct, and that the error was in Newton's drawing the figure. He suggests that Newton used discrete arcs of the circle of curvature to approximate the curve. Nauenberg then uses the method to reproduce what he argues is the correct form of Newton's figure (Nauenberg 1994, 235–239).

12. Nauenberg extends the numerical curvature method to obtain the relationship $v\partial v = -F\partial r$ and integrates it for $F = c/r^q$. The resulting expression $[2r/(q-1)] + [2E\, r^q/c]$ is equal to $\rho \sin(\alpha)$, where E is a constant of integration set by the initial conditions (i.e., the modern total energy). For Proposition 7, for example, the expression $\rho \sin(\alpha)$ is equal to $r/2$ and the expression is satisfied for $E = 0$ and the exponent $q = 5$. Thus, a solution for the problem set in Proposition 7 is $F = c/r^5$. For solutions to Propositions 9 and 11, see Nauenberg 1994, 231–232.

13. Newton [corres. 1676–1687] 1960, Vol. 2: 308.

14. Nauenberg 1994a, 234.

15. For a discussion of Bernoulli's criticism of this issue and of Newton's response to it by way of John Keill's directed correspondence, see Whiteside 1974, Vol. 6: 146 n. 124; 348 n. 209. See also Whiteside 1974, Vol. 6: 56 n. 73; 556 n. 38 for more on the inverse problem.

16. Whiteside 1974, Vol. 6: 147 n. 124.

17. Newton [corres. 1661–1727] 1975, Vol. 5: 5.

18. Newton [corres. 1718–1727] 1977, Vol. 7: 78.

19. For a presentation of the details of a potential solution of the inverse problem, see Whiteside 1974, Vol. 6: 146 n. 24.

20. The most vocal and persistent challenger is Robert Weinstock, who claims that Newton's solution is radically flawed. His first article appeared in 1982 and

bears the provocative title "Dismantling a Centuries-old Myth: Newton's *Principia* and Inverse-Square Orbits." It was followed in 1989 by "Long-buried Dismantling of a Centuries-old Myth: Newton's *Principia* and Inverse-Square Orbits." In 1991, Bruce Pourciau responded to Weinstock's challenge with an article entitled "On Newton's Proof that Inverse-Square Orbits Must Be Conics," in which he demonstrated that any flaw that existed was intuitively easy to repair. Weinstock responded in 1992 with the article "Newton's *Principia* and Inverse-Square Orbits: The Flaw Reexamined." In 1994, the *College Mathematics Journal* published a series of five articles: the first by Weinstock, entitled "Isaac Newton: Credit Where Credit Won't Do," the rest by distinguished historians of science and physicists: Anthony P. French, Michael Nauenberg, Richard S. Westfall, and Curtis Wilson. Nauenberg in particular carefully considers and counters each of the arguments put forth by Weinstock concerning the question of the inverse-square orbits. True to form, however, Weinstock refuses to concede ground to any critic. In a response to Nauenberg's reference to Proposition 17, for example, Weinstock claims that "this spurious representation of Prop. 17 has had a fascinating career, in which not only Newton, but also L. Euler and I. B. Cohen, for example, are incriminated" (Weinstock 1994, 222). Thus, Nauenberg joins the list of scholars, including myself, who, past and present, stand opposed to the claim made by Weinstock. See Nauenberg 1994b, Pourciau 1991, Weinstock 1989 and 1994.

21. Arnol'd 1990, 31–33.

22. What follows is an example that has been taken from my longer article on curvature. See Brackenridge 1992, 252–255.

23. Whiteside 1974, Vol. 6: 148 n. 124.

24. The derivative $z' = dz/d\theta$
$$= (dz/dr)(dr/d\theta)$$
$$= rz(dz/dr)$$
$$= (r/2)d(1 + z^2)/dr.$$

25. Newton [math. 1664–1666] 1967, Vol. 1: 245–297.

APPENDIX: TRANSLATION

1. I. Bernard Cohen, who translates both *corpus gyrat* and *gyretur corpus* as "a body revolves," opposes the distinction that we make (active versus passive), maintaining that to do so is to convert the rules of Latin grammar into an implied physical doctrine (an external mover), which he believes is patently wrong. He holds that it is Newton's intent to keep the discussion on a neutral plane and that it is contrary to that intent to use so positive an expression as "let a body be orbited" or "let a body be revolved" (personal correspondence). In defense of his position, he cites the long tradition of works on Newton, many produced by members of Newton's own circle, which never makes this distinction. John Clarke, the brother of Newton's colleague Samuel Clarke, translates *gyretur corpus* as if it were *corpus gyrat* "a body revolves," omitting both voice and mood changes, as does Andrew Motte in the first (1729) English translation. Cohen also points out, however, that Clarke did on occasion use the active and apparently could not decide between the two alternatives. In 1974, D. T. Whiteside evades the question of mover by reflecting the mood change (indicative to subjunctive) but not the voice change

when he translates *corpus gyrat* "a body orbits" (Newton [math. 1684–1689] 1974, Vol. 6: 42–43) and *gyretur corpus* "let a body orbit" (Newton [math. 1684–1689] 1974, Vol. 6: 550–551). Evidence in earlier astronomical texts, however, indicates that such active/passive distinctions were made. Our colleague Andrea Murschel, who is preparing a translation of Regiomontanus's *Epitome* of 1643, assures us that Regiomontanus, for example, distinguishes between *movetur* and *movet*. She writes, "Regiomontanus rarely uses the active when discussing the motion of the planets, although mathematically, an active translation of words such as *movetur* would not cloud or contradict such passages. But physically and philosophically, the presence of the passive form demonstrates without a doubt that Regiomontanus conceived of an external force propelling the planets—God the Mover. His choice of the passive was assuredly conscious" (personal correspondence). Newton did not accept the Aristotelian requirement of a constant external agent, as Regiomontanus did, but Newton was certainly aware of the subtleties of the Latin language. Cohen goes beyond the question of translation, however, and points out that when Newton wrote in English about planetary motion, he used the active voice. In the *Phaenomena*, Newton writes "The Planets Mercury & Venus revolve (active not passive voice) about the Sun" and "The Planet Mars revolves (again active) about the Sun" (Hall and Hall 1962, 383). This document is dated by the Halls as being written after the 1687 *Principia* and perhaps before the 1713 *Principia*. Thus, Cohen may well be correct in maintaining that we are converting the exigencies of Latin grammar into an implied physical doctrine beyond Newton's intent. It may well be that Newton's choice of the passive rather than the active voice in the Latin was stylistically motivated and did not indicate a conceptual position. Nevertheless, Newton did make the change and given our self-imposed constraints, we have noted that change.

2. The concerned reader is directed to the *Guide to Newton's Mathematical Principles of Natural Philosophy* by I. Bernard Cohen that accompanies the new translation of the 1726 *Principia* (Newton [prin. 1726] 1996). In this comprehensive survey of the substance as well as the surroundings of the *Principia*, Cohen brings a lifetime of Newtonian scholarship to bear on the work. In particular, he addresses the question of Newton's choice of the active versus the passive voice and offers a scholarly and detailed justification for the choice of the active voice for the new translation.

3. Our source for the Latin text of the first edition of the *Principia* is the facsimile reproduction (Newton [prin. 1687] 1953). D. T. Whiteside's translation of Newton's text in *On Motion* was of great help throughout this book (Newton [math. 1684–1691] 1974), as was the draft of I. Bernard Cohen and Ann Whitman's new translation of the 1726 *Principia* (Newton [prin. 1726] 1996) made available to us by I. Bernard Cohen.

REFERENCES

PRIMARY WORKS

Newton, Isaac. 1684. *De Motu Corporum in Gyrum* (On the motion of bodies in orbit). *Register Book of the Royal Society* 6: 218–234.

———. [1684–1685] 1989. *The Preliminary Manuscripts for Isaac Newton's 1687 Principia 1684–1685*. Ed. D. T. Whiteside. Cambridge: Cambridge University Press.

———. [corres. 1661–1727] 1959–1977. *The Correspondence of Isaac Newton*. 7 vols. Trans. and ed. H. W. Turnbull (vols. 1–3), J. F. Scott (vol. 4), A. R. Hall and Laura Tilling (vols. 5–7). Cambridge: Cambridge University Press.

———. [math. 1670–1722] 1967–1981. *The Mathematical Papers of Isaac Newton*. 8 vols. Trans. and ed. D. T. Whiteside. Cambridge: Cambridge University Press. (Newton's text is cited as "Newton" but the extensive notes by the editor in these volumes are cited as "Whiteside.")

———. [prin. 1687] 1953. *Philosophiae Naturalis Principia Mathematica*. 1st ed. Facsimile ed. London: William Dawson and Sons.

———. [prin. 1726] [1729] 1968. *The Mathematical Principles of Natural Philosophy*. 3d ed. 2 vols. Trans. Andrew Motte. Facsimile ed. London: William Dawson and Sons.

———. [prin. 1726] 1972. *Philosophiae Naturalis Principia Mathematica*. 3d ed., with variant readings. 2 vols. Eds. Alexandre Koyré and I. Bernard Cohen. Cambridge: Cambridge University Press.

———. [prin. 1726] 1996. *Mathematical Principles of Natural Philosophy*. 3d ed. Trans. I. Bernard Cohen and Anne Whitman, assisted by Julis Budenz. With an introductory "Guide to Newton's *Mathematical Principles of Natural Philosophy*," by I. Bernard Cohen. Berkeley: University of California Press.

SECONDARY WORKS

Apollonius. [c. 200 B.C.] 1952. *On Conic Sections*. Trans. R. Catesby Taliaferro (1939). In *Great Books of the Western World*, vol. 11, ed. Robert Maynard Hutchins. Chicago: Encyclopaedia Britannica.

Aristotle. [c. 350 B.C.] 1961. *Aristotle's Physics.* Trans. Richard Hope. Lincoln: University of Nebraska Press.

Arnol'd, V. I. 1990. *Huygens and Barrow, Newton and Hooke: Pioneers in Mathematical Analysis and Catastrophe Theory form Evolvents to Quasicrystals.* Trans. Eric J. F. Primrose. Basel: Birkhäuser Verlag.

Ball, W. W. Rouse. 1893. *An Essay on Newton's* Principia. London: Macmillan.

Barbour, Julian B. 1989. *Absolute or Relative Motion? A Study from a Machian Point of View of the Discovery and the Structure of Dynamical Theories.* Vol. 1, *The Discovery of Dynamics.* Cambridge: Cambridge University Press.

Bertoloni Meli, Domenico. 1990. "The Relativization of Centrifugal Force." *Isis* 81: 23–43.

———. 1993. *Equivalence and Priority: Newton versus Leibniz.* Oxford: Clarendon Press.

Blake, William. [1789] 1975. *The Marriage of Heaven and Hell.* Ed. Geoffrey Keynes. New York: Oxford Press.

Brackenridge, J. Bruce. 1982. "Kepler, Elliptical Orbits, and Celestial Circularity: A Study in the Persistence of Metaphysical Commitment." *Annals of Science* 39: 117–143, 265–295.

———. 1985a. "The Defective Diagram as an Analytical Device in Newton's *Principia.*" In *Religion, Science and Worldview: Essays in Honor of Richard S. Westfall,* eds. M. J. Osler and P. L. Farber. Cambridge: Cambridge University Press.

———. 1985b. "Kuhn, Paradigms, and Astronomy: Astronomy as a Case Study of Kuhnian Paradigms." *Proceedings of the American Philosophical Society* 129: 433–455.

———. 1988. "Newton's Mature Dynamics: Revolutionary or Reactionary?" *Annals of Science* 45: 451–476.

———. 1990. "Newton's Unpublished Mature Dynamics: A Study in Simplicity." *Annals of Science* 47: 3–31.

———. 1992. "The Critical Role of Curvature in Newton's Developing Dynamics." In *An Investigation of Difficult Things: Essays on Newton and the History of the Exact Sciences,* eds. P. M. Harman and Alan E. Shapiro. Cambridge: Cambridge University Press.

———. 1993. "The Locke/Newton Manuscript: Conjugates, Curvatures, and Conjectures." *Archives internationales d'histoire des sciences* 43: 280–292.

Brackenridge, J. Bruce, and Robert M. Rosenberg. 1970. *The Principles of Physics and Chemistry.* New York: McGraw-Hill.

Brackenridge, J. Bruce, and Mary Ann Rossi. 1979. "Johannes Kepler's 'On the More Certain Fundamentals of Astrology.'" *Proceedings of the American Philosophical Society* 123: 85–116.

Chandrasekhar, S. 1995. *Newton's* Principia *for the Common Reader.* Oxford: Clarendon Press.

Clarke, John. [1730] 1972. *A Demonstration of Some of the Principal Sections of Sir Isaac Newton's* Principles of Natural Philosophy. New York: Johnson Reprint Corporation.

Cohen, I. Bernard. 1970. "Newton's Second Law and the Concept of Force in the *Principia.*" In *The* Annus Mirabilis *of Sir Isaac Newton 1666–1966,* ed. Robert Palter. Cambridge: M.I.T. Press.

———. 1971. *Introduction to Newton's* Principia. Cambridge: Cambridge University Press.

———. 1983. *The Newtonian Revolution.* Rev. ed. Cambridge: Cambridge University Press.
———. 1985. *The Birth of a New Physics.* Rev. ed. New York: W. W. Norton.
Cook, Alan. 1991. "Edmund Halley and Newton's *Principia.*" *Notes and Records of the Royal Society* 45: 129–134.
Dampier, W. C. 1966. *A History of Science and Its Relations with Philosophy and Religion.* 4th ed., with a postscript by I. Bernard Cohen. Cambridge: Cambridge University Press.
Descartes, René. [1644] 1983. *Principles of Philosophy.* Trans. Valentine Rodger Miller and Reese P. Miller. Dordrecht: D. Reidel.
Dobbs, Betty Jo Teeter. 1991. *The Janus Faces of Genius: The Role of Alchemy in Newton's Thought.* Cambridge: Cambridge University Press.
Erlichson, Herman. 1991. "Motive Force and Centripetal Force in Newton's Mechanics." *American Journal of Physics* 59: 842–849.
———. 1992a. "The Instantaneous Impulse Construction as a Formula for Central Force Motion on an Arbitrary Plane Curve with Respect to an Arbitrary Force Center in the Plane of that Curve." *Annals of Science* 49: 369–375.
———. 1992b. "Newton's Solution to the Equiangular Spiral Problem and a New Solution Using only the Equiangular Property." *Historia Mathematica* 19: 402–413.
———. 1992c. "Newton's Polygonal Model and the Second Order Fallacy." *Centarurus* 35: 243–258.
———. 1993. "The Riddle of the Kepler-Motion Papers." *Archives internationales d'histoire des sciences* 43: 258–279.
Euclid. [c. 300 B.C.] 1952. *The Thirteen Books of Euclid's Elements.* Trans. R. Catesby Taliaferro (1939). In *Great Books of the Western World,* vol. 11, ed. Robert Maynard Hutchins. Chicago: Encyclopaedia Britannica.
Galileo. [1638] 1954. *Dialogues Concerning Two New Sciences.* Trans. H. Crew and A. de Salvio. New York: Dover Publications.
Garisto, Robert. 1991. "An Error in Isaac Newton's Determination of Planetary Properties." *American Journal of Physics* 59: 42–48.
Hall, A. Rupert. 1957. "Newton on the Calculation of Central Forces." *Annals of Science* 13: 62–71.
Hall, A. Rupert, and Marie Boas Hall. 1962. *Unpublished Scientific Papers of Isaac Newton.* Cambridge: Cambridge University Press.
———. 1963. "The Dates of 'On Motion in Ellipses.'" *Archives internationales d'histoire des sciences* 16: 23–28.
Herivel, John W. 1961. "Newtonian Studies III. The Originals of the Two Propositions Discovered by Newton in December 1679." *Archives internationales d'histoire des sciences* 14: 23–33.
———. 1963. "Newtonian Studies IV." *Archives internationales d'histoire des sciences* 16: 14.
———. 1965. *The Background to Newton's* Principia*: A Study of Newton's Dynamical Researches in the Years 1664–84.* Oxford: Clarendon Press.
Hooke, Robert. [1679] 1945. "Lectiones Cutlerianae." *Early Science in Oxford,* vol. 8, ed. R. T. Gunther. Oxford: Oxford University Press.
Kepler. [1619] 1952. *The Harmonies of the World: Book 5.* In *Great Books of the Western World,* vol. 16, ed. Robert Maynard Hutchins. Chicago: Encyclopaedia Britannica.

Koyré, Alexandre. [1965] 1968. *Newtonian Studies.* Chicago: University of Chicago Press.

Lamb, H. 1923. *Dynamics.* Cambridge: Cambridge University Press.

Nauenberg, M. 1994a. "Newton's Early Computational Method for Dynamics." *Archive for History of Exact Sciences* 46: 221–252.

―――. 1994b. "Newton's *Principia* and Inverse-Square Orbits." *College Mathematics Journal* 25: 212–219.

Osgood, W. F. 1937. *Mechanics.* New York: Macmillan.

Plato. [c. 350 B.C.] 1965. *Timaeus and Critias.* Trans. H. D. P. Lee. New York: Penguin.

Pourciau, Bruce. 1991. "On Newton's Proof that Inverse-Square Orbits Must Be Conics." *Annals of Science* 48: 159–172.

Symon, Keith R. 1971. *Mechanics.* 3d ed. Cambridge, Mass.: Addison-Wesley.

Thoren, Victor E. 1974. "Kepler's Second Law in England." *British Journal for the History of Science* 7: 243–256.

Wallace, W. A. 1988. "Newton's Early Writings: Beginnings of a New Direction." In *Newton and the New Direction in Science,* eds. G. V. Coyne, M. Heller, and J. Zycinski. Proceedings of the Cracow Conference. Vatican City State: Libreria Editrice Vaticana.

Weinstock, Robert. 1989. "Long-Buried Dismantling of a Centuries-old Myth: Newton's *Principia* and Inverse-Square Orbits." *American Journal of Physics* 58: 846–849.

―――. 1994. "Isaac Newton: Credit Where Credit Won't Do." *College Mathematics Journal* 25: 179–192.

Westfall, R. S. 1963. "A Note on Newton's Demonstration of Motion in Ellipses." *Archives internationales d'histoire des sciences* 22: 51–60.

―――. 1980. *Never at Rest: A Biography of Isaac Newton.* Cambridge: Cambridge University Press.

Whiteside, D. T. 1964. "Newton's Early Thoughts on Planetary Motion: A Fresh Look." *British Journal of the History of Science* 2: 117–137.

―――. 1966. "Newtonian Dynamics." *History of Science* 5: 104–177.

Whiteside, D. T., ed. 1967. *The Mathematical Papers of Isaac Newton.* Vol. 1, *1664–1666*. With introduction and notes by the editor. Cambridge: Cambridge University Press. (Newton's text is cited as "Newton, Isaac. [math. . . .]" but the extensive notes by the editor in these volumes are cited as "Whiteside.")

―――. 1969. *The Mathematical Papers of Isaac Newton.* Vol. 3, *1670–1673*. With introduction and notes by the editor. Cambridge: Cambridge University Press.

―――. 1970. "The Mathematical Principles Underlying Newton's *Principia Mathematica.*" *Journal for the History of Astronomy* 1: 116–138.

―――. 1972. *The Mathematical Papers of Isaac Newton.* Vol. 5, *1683–1684*. With introduction and notes by the editor. Cambridge: Cambridge University Press.

―――. 1974. *The Mathematical Papers of Isaac Newton.* Vol. 6, *1684–1691*. With introduction and notes by the editor. Cambridge: Cambridge University Press.

―――. 1989. *The Preliminary Manuscripts for Isaac Newton's 1687 Principia 1684–1685*. With introduction and notes by the editor. Cambridge: Cambridge University Press.

Yoder, Joella G. 1988. *Unrolling Time: Christiaan Huygens and the Mathematization of Nature.* Cambridge: Cambridge University Press.

INDEX TO THE GUIDED STUDY AND THE TRANSLATION

UNIFORM CIRCULAR MOTION

Circular Motion (*Waste Book*)	guided study, 45–54; polygonal approximation, 42–45
On Circular Motion	guided study, 58–63; parabolic approximation, 56–58
Proposition 4 (*Principia*)	an alternate demonstration, 54–56

ON MOTION (ON MOTION OF BODIES IN ORBIT)

Definition 1	discussion, 74; and Definition 5 (*Principia*), 143
Definition 2	discussion, 75; and Definition 3 (*Principia*), 143
Definition 3	discussion, 75
Hypothesis 1	discussion, 75
Hypothesis 2	discussion, 75; and Law 1 (*Principia*), 146
Hypothesis 3	discussion, 75; revision, 77–79; and Corollary 1 (*Principia*), 146–147
Hypothesis 4	discussion, 76; and Lemma 10 (*Principia*), 150
Lemma 1	discussion, 76; and Lemma 12 (*Principia*), 161
Lemma 2	discussion, 76
Problem 1	guided study, 97–102; and Proposition 7 (*Principia*), 157
Problem 2	guided study, 102–106; and Proposition 10 (*Principia*), 161–162
Problem 3	guided study, 106–117; and Proposition 11 (*Principia*), 162–163
Problem 4	guided study, 125–135; and Proposition 17 (*Principia*), 164–165
Problem 5	73
Problem 6	73
Problem 7	73
Theorem 1	guided study, 79–85; and Proposition 1 (*Principia*), 154–155
Theorem 2	guided study, 85–90; and Proposition 4 (*Principia*), 155–156
Theorem 3	guided study, 90–93; and Proposition 6 (*Principia*), 156–157
Theorem 4	guided study, 119–125; and Proposition 15 (*Principia*), 164

PRINCIPIA (THE MATHEMATICAL PRINCIPLES OF NATURAL PHILOSOPHY)

Book One

Dedicatory poem	translation, 233–234
Definition 1	143
Definition 2	143
Definition 3	and Definition 2 (*On Motion*), 143
Definition 4	143
Definition 5	and Definition 1 (*On Motion*), 143
Definition 6	discussion, 143
Definition 7	discussion, 143
Definition 8	discussion, 144; and Law 2, 146
Law 1	and Hypothesis 2 (*On Motion*), 146
Law 2	discussion, 146
Law 3	146
Lemma 1	discussion, 148–149; translation, 235
Lemma 2	translation, 235–236
Lemma 3	translation, 236
Lemma 4	translation, 237
Lemma 5	translation, 237–238
Lemma 6	translation, 238
Lemma 7	translation, 238–239
Lemma 8	translation, 239–240
Lemma 9	translation, 240
Lemma 10	and Hypothesis 1 (*On Motion*), 150; translation, 241
Lemma 11	discussion, 150–154; revision, 184–187; translation, 241–144
Lemma 12	and Lemma 1 (*On Motion*), 161; translation, 255
Lemma 13	translation, 260
Lemma 14	translation, 260–261
Preface	translation, 230–232
Proposition 1	revised, 187–190; and Theorem 1 (*On Motion*), 154–155; translation, 245
Proposition 2	discussion, 155; translation, 246–247
Proposition 3	discussion, 155; translation, 247–248
Proposition 4	scholium, 54–55; and Theorem 2 (*On Motion*), 155–156; translation, 248–250
Proposition 5	discussion, 156; translation, 251
Proposition 6	and the orbital equation, 215–216; rejected revision, 196–198; revised, 190–196; and Theorem 3 (*On Motion*), 156–157; translation, 252
Proposition 7	and the orbital equation, 217; and Problem 1 (*On Motion*), 157; revised, 198–201; translation, 252–253
Proposition 8	discussion, 157; translation, 253–254
Proposition 9	discussion, 157–161; and the orbital equation, 212, 217; revised, 202–204; translation, 254–255
Proposition 10	and the orbital equation, 217; and Problem 2 (*On Motion*), 161–162; revised, 204–205; translation, 255–256
Proposition 11	and the orbital equation, 212, 217; and Problem 3 (*On Motion*), 162–163; revised, 205–209; translation, 257–258
Proposition 12	discussion, 163; translation, 258–260
Proposition 13	Corollary 1, 163–164, 218–220; discussion, 163; translation, 261–262

Proposition 14	discussion, 164; translation, 262–263
Proposition 15	and Theorem 4 (*On Motion*), 164; translation, 263
Proposition 16	discussion, 164; translation, 263–265
Proposition 17	and Problem 4 (*On Motion*), 164–165; translation, 265–267

Book Three

Proposition 13	discussion, 135–136

Proposed Radical Revisions

Proposed Lemma 12	discussion, 175; partial text, 282
Proposed Proposition 6	partial text, 281; and the *Principia*, 168
Proposed Proposition 7	partial text, 281; and the *Principia*, 168
Proposed Proposition 8	comparison theorem, 173–175; partial text, 281; and the *Principia*, 168–169; and Proposition 7 (Corollary 3), 201
Proposed Proposition 9	circular dynamics ratio, 171, 175; linear dynamics ratio, 171; partial text, 281–282; and the *Principia*, 169
Proposed Proposition 10	applications, 175–176; partial text, 282; and the *Principia*, 169
Proposed Proposition 11	169
Proposed Proposition 12	169
Proposed Proposition 13	169–170

GENERAL INDEX

Apollonius: *latus rectum*, 112, 119–120; prop. 15 (bk. I), 112, 114, 194; prop. 48 (bk. III), 111, 129; prop. 31 (bk. VII), 76, 161
Apple story. *See* Newton, Isaac
Area law: conservation of angular momentum, 213; Kepler, 6–7; polygonal approximation, 24–27; Proposition 1, 154–155; rejection of celestial vortices, 23–24; revision of Proposition 1, 187–190; Theorem 1, 79–85
Aristotle: circular motion, 14; falling bodies, 14; *Meteorology*, 41; *On Generation and Corruption*, 41; *On the Heavens*, 41; *Physics*, 14, 17–18, 41, 145; rectilinear motion, 17–18; space and time, 145
Arnol'd, V.I., 220

Barbour, Julian, 39, 145
Barrow, Isaac, 42
Bentley, Richard, 3
Bernoulli, Johann, 126, 218–219
Bertoloni Meli, Dominico, 38, 126
Blake, William, 10–11
Boyle, Robert, 41
Brahe, Tycho, 16, 117

Calculus. *See* Fluxions
Calendar (old style / new style), 273
Cambridge University, 13, 41
Centrifugal force. *See* Force
Centripetal force. *See* Force

Circle of curvature. *See* Curvature
Circular approximation, 171, 222; circular dynamics ratio, 35–37; elliptical motion, 63–65; Lemma 11 (Corollary 3), 187; orbital equation, 213–214; Proposition 7 (Corollary 3), 201; Whiteside, 187. *See also* Curvature
Circular dynamics ratio: circular approximation, 35–37, 171; curvature, 182–183; orbital equation, 216, 222; revised Proposition 6, 191–193
Circular motion. *See entry in the Index to the Guided Study and the Translation*
Clarke, John, 280, 286
Cohen, I. Bernard, xii, 38, 146, 286–287
Comets, 132–134
Comparison theorem: Proposition 8 (proposed), 172–175; Proposition 7, 201; Proposition 11 (alternate method), 207–208
Conduitt, John, 5
Conjugate diameters: construction, 103–104; lost solution of 1679, 177–180
Copernicus, 12, 14, 24
"Crookednesse": elliptical motion, 63; Locke solution, 179. *See also* Curvature
Curvature: circle of, 63–65; circular dynamics ratio, 182–183; Lemma 11, 150–155; Lemma 11 (revised), 184–187; orbital equation, 213–215; Proposition 7 (Corollary 3), 201. *See also* Circular approximation

297

GENERAL INDEX

Curvature orbital equation. *See* Orbital equation

Demoivre, Abraham: Halley's visit to Newton, 71; lost solution of 1679, 177
Descartes, René, 12–13, 15; ball in a sling and tube, 19; change in motion, 18; *Geometry*, 13, 41–42; influence on Newton, 17–24; outward endeavor, 21–22, 45, 82, 86; *Philosophical Principles*, 12–13; *Principia*, 142; uniform circular motion, 18; uniform rectilinear motion, 17; vortex, 12–13, 15, 22–23, 142
Direct problem: defined, 15–17; paradigm, 33, 96–97. *See also* Kepler problem
Discriminate ratio: defined, 34–35; Problem 1, 97–100; Problem 2, 104–105; Problem 3, 107, 109–110
Dobbs, Betty Jo Teeter, ix; area law and celestial vortices, 23; spiritual cause for gravity, 270–271

Einstein, Albert, 145
Erlichson, Herman, xii, 271, 280, 283
Euclid: prop. 13 (bk. II), 130; prop. 31 (bk. III), 151–152; prop. 32 (bk. III), 278; prop. 35 (bk. III), 194–196; prop. 36 (bk. III), 60–61, 87–88, 100
Euler, Leonhard: Proposition 17 and the inverse problem, 126, 278

Flamsteed, John, 69–70
Fluxions: early dynamics, 42, 176; *Methods of Series and Fluxions*, 42, 183, 217
Force: centrifugal, 19–22, 38, 55; centripetal, 20, 59, 82–83, 143–144, 150; gravitational, vii, 6, 10, 73; impulsive, 18, 38, 50–54, 274; innate, 75, 146; reflection, 46, 51. *See also* Newton's second law

Galileo, 147, 256; inclined planes, 8, 15; influence on Newton, 27–29; projectile motion, 31; *Two New Sciences*, 14–15, 41
Grantham, England, 40
Gregory, David: on Newton's proposed revisions, 166–167, 170, 172

Halley, Edmund: dedicatory poem, 142–143, 233–234; *On Motion*, 70–74, 136–137; planetary positions, 124–125
Herivel, John, 273, 275, 282

Hobbes, Thomas, 41
Hooke, Robert: and Newton's dynamics, 20–24, 69–70, 276
Huygens, Christiaan: centrifugal force, 19–20, 176; *Principia*, 9–10, 147

Impulse. *See* Force
Inertia: Aristotle, 17–18; Descartes, 15, 17, 29, 43; Newton, 17, 29, 43, 146
Inverse problem, 15–16. *See also* Kepler problem: inverse solution

Jupiter. *See* Planets and satellites

Keill, John, 126
Kepler, Johannes, vii; celestial circularity, 12; *New Astronomy*, 12; *World Harmony*, 90
Kepler problem: description, vii, 15–16; direct solution, 5–9, 33–35, 106–117, 205–209, 212; inverse solution, 15, 70, 126, 218–221. *See also* Locke solution

Latus rectum. *See* Apollonius
Leibniz, Wilhelm Gottfried, 9–10, 70
Linear dynamics ratio: definition, 7–8; orbital equation, 215–216; parabolic approximation, 29–33; Proposition 6, 156–157; Proposition 6 (revised), 190–191; Proposition 9 (proposed), 171; Theorem 3, 90–93, 222
Locke, John, 10, 65, 176
Locke solution, 176–180
Lost solution of 1679, 6; Demoivre's statement, 71. *See also* Locke solution

Mars. *See* Planets and satellites
Mechanical philosophy, 23, 41
Mercury. *See* Planets and satellites
Methods of Series and Fluxions. *See* Fluxions
Moon. *See* Planets and satellites
Murschel, Andrea, xii, 287

Nauenberg, Michael, vi, 217–218, 285, 286
Newton, Isaac: advice to readers, 3–5; apple story, 5–6; Blake's portrait of, 10–11; Descartes, 17–24; early life, 40–42; Galileo, 27–29; Halley, 70–74, 136–137; Hooke, 20–24, 69–70, 276. See also *On Motion*; *Principia*
Newton's second law, 37–38, 144–146, 211

On Circular Motion, 58–63; parabolic approximation, 56–58
On Motion (*On Motion of Bodies in Orbit*): description, 71–74; relationship to *Principia*, 4, 141–142, 227; sent to Halley, 71–74. *See also entry in the Index to the Guided Study and the Translation*
On Motion in Ellipses. *See* Locke solution
On Motion of Spherical Bodies in Fluids, 77–79
Orbital equation: curvature orbital equation, 213–215, 220–221; curvature orbital measure, 216–217; polar orbital equation, 212
Osculating circle. *See* Circle of curvature
Osgood, William Fogg, 19
Outward endeavor. *See* Descartes

Paolozzi, Eduardo, 269
Pappus, 230
Parabolic approximation: definition, 28; elliptical orbits, 33–35; linear dynamics ratio, 29–33; uniform circular motion, 56–57; Whiteside, 31
Parallelogram rule, 28; combining forces, 37–38; continuous force, 79; impulsive force, 77–79; Proposition 1 (revised), 187–189; virtual displacements, 49
Pepys, Samuel, 142, 229
Planets and satellites: earth's moon, 5, 58; Jupiter, 90, 124, 135–136; Jupiter's moons, 90, 135, 142; Mars, 16, 35, 106, 117, 124; Mercury, 124–125; Saturn, 90, 124, 135–136; Saturn's moons, 90; Venus, 124–125, 135
Plato, 12, 145
Polar orbital equation. *See* Orbital equation
Polygonal approximation: area law, 24–27, 79–80; uniform circular motion, 42–43
Pope, Alexander, 10–11
Pourciau, Bruce, xi, 286
Principia (*The Mathematical Principles of Natural Philosophy*): description, 141–143; proposed revisions, 4–5, 166–168, 170–173; publication, 3–4, 136–137; published revisions, 4–5, 182–184, 209–210; reception of, 9–11; relationship to *On Motion*, 4, 141–142. *See also entries in the Index to the Guided Study and the Translation*

Problem of the planets, 12–14
Ptolemy, 12, 24

Register Book. *See* Royal Society
Royal Society, xi; Halley and, 68, 136; Hooke and, 69–70; Newton and, 2, 140; *On Motion* and, 73–74, 77, 136; *Principia* and, 10, 74, 137, 231; *Register Book*, 74, 136; Somerville and, 224

St. Helena, 125
Saturn. *See* Planets and satellites
Second law. *See* Newton's second law
Shapiro, Alan, xii
Somerville, Mary, 223–224
Space: absolute, 144–146; Aristotle, 145

Time: absolute, 144–146; Aristotle, 145; particles of, 38, 146; Plato, 145
Trinity College, Cambridge, 41, 142
Twain, Mark, 3

Uniform circular motion. *See entries in the Index to the Guided Study and the Translation*
Uniqueness. *See* Kepler problem: inverse solution

Venus. *See* Planets and satellites
Virtual displacement, 27, 49–50
Voltaire, François, 10
Vortex. *See* Descartes

Wallis, John, 147
Waste Book: description, 40, 42–44; uniform circular motion, 45–54
Weinstock, Robert, 278, 285–286
Westfall, Richard, 20, 37, 269, 286
Whiteside, D. T., vi; circular approximation, 187; Hooke, 24; inverse problem, 219–220; lemmas, 1–11, 148; orbital equation, 215; parabolic approximation, 31; radical revisions, 166–168, 171
Woolsthorpe, England, 40
Wren, Christopher, 71, 118, 147, 250

Designer:	UC Press Staff
Compositor:	Prestige Typography
Text:	10/12 Baskerville
Display:	Baskerville
Printer:	Bookcrafters, Inc.
Binder:	Bookcrafters, Inc.

QB 355 .B694 1995 c.1
Brackenridge, J. Bruce, 1927—

The key to Newton's dynamics

DATE DUE

GAYLORD   PRINTED IN U.S.A.